High Performance
Memories

High Performance Memories

New architecture DRAMs and SRAMs – evolution and function
Revised Edition

Betty Prince
Memory Strategies International
Leander, Texas
USA

JOHN WILEY & SONS, LTD
Chichester · New York · Weinheim · Brisbane · Singapore · Toronto

Contents

Acknowledgments

I would like to thank all of those who contributed information and offered suggestions for this book. First I would like to thank my husband, Joseph Hartigan, who contributed both extreme patience during the writing of the book and his own expertise in the memory applications area gained as Senior Systems Engineer in the DRAM group at Texas Instruments, and currently as Senior Engineer at Toshiba.

Other experts to whom I owe much gratitude for the time and care they took in reading and making suggestions on various parts of this book include:

Mark Nishiwaki of Oki Semiconductor and his DRAM design people, and Cormac O'Connell, Design Manager at Mosaid who read and offered excellent suggestions on the Asynchronous DRAM chapter.

Fred Jones, Design Manager, and Michael Parris, Senior Designer, at United Memories, and Sam Chen, Applications Manager for DRAMs at Mitsubishi, for their invaluable comments on the Synchronous DRAM chapter.

Mike Peters, Fast SRAM Applications Manager at Motorola, for sending material to be used in the book and for offering excellent suggestions for the two SRAM chapters.

Dick Lawrence, Systems Engineer at Digital Equipment, and Cormac O'Connell, of Mosaid, for reading and for many helpful comments on the Graphics DRAM chapter.

Bruce Wenniger, with the Application Group at Cypress, for reading and commenting extensively on the Interface chapter and the Cache SRAM chapter, and Jeff Linden, previously Director of SRAM Design at Cypress, for contributing the photograph of the 1M Burst SSRAM.

Joseph Hartigan, for reading and commenting on the applications and packaging chapters, and Derek Best, of Micromodule Systems, for sending information and photographs for the packaging chapter.

Finally, I would like to thank all the people in the various memory companies who spent time getting permissions from their management for me to use material from their company's databooks.

About the author

Betty Prince is President of Memory Strategies International – a semiconductor memory services company. She has a 25-year history in the semiconductor industry, having held management positions in development engineering, marketing, and operations at Fairchild, Motorola, N.V. Philips, and most recently Texas Instruments. Dr Prince is author of the book, *Semiconductor Memories* (Wiley, 1991). She has been involved in the EIA JEDEC JC42 Memory Standards Committee since 1982, was chairman of the JC16 Electrical Interface Standards Committee (1991–92), and is currently a US National Delegate to the IEC SC47A WG3 International Memory Standards Committee. She has been on the Board of Directors of Mosaid Technologies since 1996, on the Scientific Advisory Board of Silicon Access since 1997, on the Editorial Board of the IEEE Spectrum from 1991 to 1994 and has been Keynote Speaker on Memories at several IEEE Conferences. She has a BS and MS in Physics from the Universities of New Mexico and California, an MBA in International Business and a PhD in International Finance, both from the University of Texas, Austin. Her doctoral dissertation was on Chaotic Attractor Modeling (1987).

Introduction

The goal of this book is to provide an overview of the major products and trends in the rapidly developing field of high performance semiconductor memories. Chapter 1 is an overview of the book. Chapter 2 discusses the systems environment which requires the high performance of memories. Chapters 3 through 7 are product specific, covering the various categories of high speed SRAMs and DRAMs. Chapter 3 discusses various categories of high speed SRAM including latched, registered, wide and separate I/O, dual port, and FIFOs. Chapter 4 discusses elementary cache and high performance cache SRAMs, Chapter 5 explores fast asynchronous DRAMs, Chapter 6 discusses fast synchronous and new architecture DRAMs, and Chapter 7 discusses DRAMs targeted at graphics applications. Chapter 8 discusses electrical characteristics of high speed memories, including interfaces, system characteristics, and test. Chapter 9 discusses the effect of packaging on memory speed.

Disclaimer

This book aims to provide accurate and authoritative information in the area of advanced memory design. Readers are, however, advised to obtain the latest information from the organisation in question. The author and the publisher specifically disclaim any and all liability arising directly from acting or failing to act on any information contained in this book.

1 Overview of High Speed Memories and Memory Systems

1.1 Overview of Fast Memory Trends

More performance is being demanded of memories by processors in high powered PCs and workstations, as well as by cache and graphics subsystems and high speed communications equipment. Memory manufacturers are responding with an array of innovative fast memory chips for the various applications.

While both DRAMs and SRAMs are making significant gains in speed and bandwidth, there still remains a gap with the speed requirements of the processors as shown in Figure 1.1. Processor internal speeds are running up to 500 MHz in 1998 and are expected to reach 800 MHz or higher by the end of the decade. The fastest RAM is only expected to reach a 250 MHz clock cycle in this time period.

The solution for providing adequate memory bandwidth depends on the system architecture, the application requirements and the processor all of which help determine the memory type. Limitations on speed include delays in the chip, the package, and the system.

The bandwidth mismatch is worse between the microprocessors and the DRAMs than the SRAMs, because the cycle time of a SRAM is inherently faster than that of a DRAM.

An SRAM stores data in a flipflop consisting of normal logic transistors whereas the DRAM cell uses a capacitor to store the data. The SRAM is ready for another read cycle as soon as the first is complete. The destructive read-out of the DRAM capacitor cell results in a longer cycle time than the SRAM's due to the need to write the data back into the cell after a read access and to perform precharge before the next attempt to read the device.

Systems designers can not always switch to SRAMs for their fast memory needs since SRAMs are a quarter the capacity of DRAMs for the same process technology and chip size. As the cost of a memory is determined in large part by the chip size, a SRAM tends to cost about four times as much per bit as a DRAM.

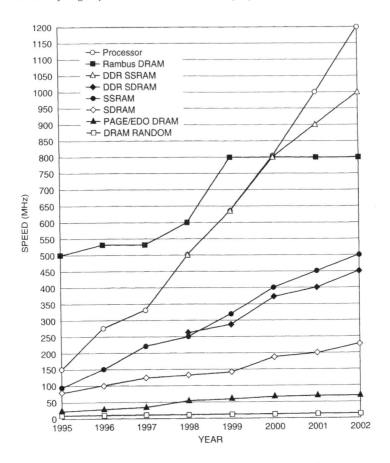

Figure 1.1 Processor and RAM speed trends

Since the main justification for the existence of the DRAM is its lower cost, the primary focus in the DRAM industry is on cost reduction. The cost and density disparity between the two devices keeps the DRAMs in demand. A faster DRAM or cheaper SRAM could eliminate the expense of using both parts in some systems.

This cost disparity with DRAMs has encouraged SRAM suppliers to offer applications-specific SRAMs which improve performance in the cache and buffer parts of systems. The higher performance justifies the higher cost per bit of the SRAM.

The Fast SRAM roadmap in Figure 1.2 indicates some of the factors in the increasing performance of the SRAMs. Faster processors, higher bus speeds and wider buses require higher bandwidth from the cache SRAMs. Cache size is increasing from 128KB (kilobytes) in the early 1990s to 512KB for a PC system by the end of the decade.

A rapidly growing classification of fast memory architecture is the Synchronous RAM. The burst synchronous SRAMs, however, have set new records for fast SRAM sales due to their inclusion in the cache in mainstream PC systems. Pipelining has also been used for high speed systems and double data rate (DDR) SRAMs, which handle two words of data per clock cycle, have appeared. DRAMs with fast cores, such as multibank DRAMs and integrated cache DRAMs, have also competed in fast

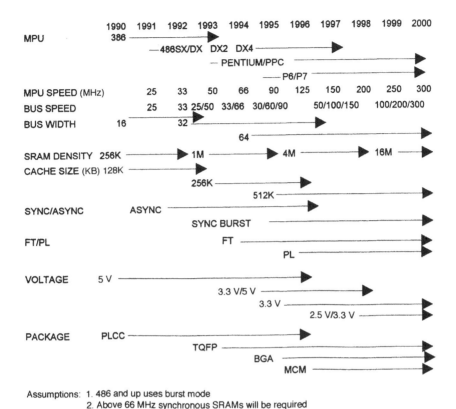

Figure 1.2 Fast SRAM roadmap (adapted from Motorola)

SRAM applications. Lower voltages have helped reduce the power dissipation of the fast SRAMs, and the reduced inductance of miniature packages also enhances the SRAM speed.

DRAM vendors are also seeking higher speed for main memory and have upgraded the old standby Fast Page Mode in the asynchronous DRAMs to a new faster version called Hyperpage Mode or extended data out, and beyond that have added synchronous interfaces and burst modes.

Synchronous DRAMs are also becoming more common and many versions are being developed. These include Cache DRAM, SDRAM, DDR SDRAM, Virtual Channel SDRAM, Enhanced ESDRAM, and DDR ESDRAM, which are intended primarily for main memory, and a variety of parts intended for multimedia and graphics subsystems, such as the synchronous Graphics DRAM which is a variant of the JEDEC SDRAM, and the DRAMs developed by Rambus, Inc. and Mosys, Inc.

1.2 New Memory Architectures to Improve Bandwidth

There are many ways to improve the bandwidth of a RAM including dividing up the internal architecture, using wider buses, using alternative output modes which access more rapidly, and using alternative technologies to MOS such as bipolar and GaAs.

A divided architecture can result in the actual datapath for the specific memory cells accessed being shorter and in the capacitive load being less so that the device switches faster. The penalty is that a divided architecture requires additional wire routing which means either more silicon or more layers for interconnects on the silicon. As the density, chip size, and bus width of DRAMs and SRAMs increase, multiply divided architectures are becoming common. It is not unusual for a standard RAM to contain 16 to 64 major internal divisions, as shown in the die photo in Figure 1.3 of a 64K×18 burst synchronous SRAM from Cypress Semiconductor.

The divisions in a standard RAM are all one logical bank. Some new architectures, however, feature independent banks on a single chip. Each small bank is faster than a

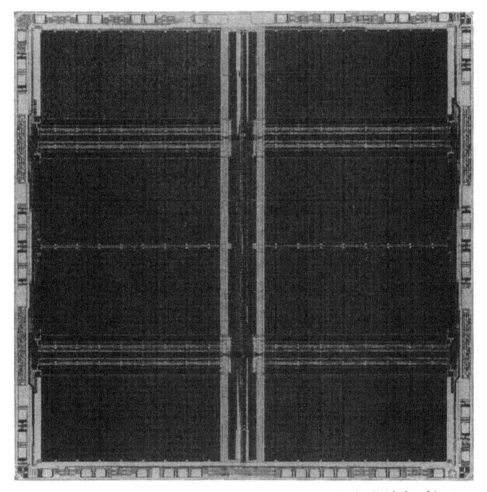

Figure 1.3   Die photograph of 64K×18 SSRAM showing multiply divided architecture
(source: Cypress Semiconductor)

single large bank would have been. Synchronous DRAMs, DRAMs from Rambus, Inc. and Mosys Inc., are a few of the parts that use small independent banks on a chip to gain speed. A block diagram of a two-bank synchronous DRAM is shown in Figure 1.4.

Wider buses are used to improve the bandwidth which is the rate at which data can be moved into or out of the memory. A RAM with a 32-bit wide bus clearly has 32 times the data throughput rate of a RAM with a 1-bit wide bus.

The limitations to bus width are package size, board space for the external traces, and ringing due to high peak currents when the wide outputs swing. All of these limitations are currently being addressed. Miniature packages and small memory modules address the package size and board space issue. New low voltage operating ranges and low swing interfaces are addressing the peak current issue.

Another method of speeding up the memory is to use alternative technologies such as bipolar or GaAs. It is not uncommon in CMOS circuits for elements of the architecture requiring high drive current such as outputs and line drivers to be made of bipolar circuit elements, a technology called BiCMOS.

The disadvantage of BiCMOS is higher standby power consumption, larger silicon area and a more expensive process resulting in higher cost. In systems where speed in the GHz range is required, and where the expense of appropriate cooling techniques is not prohibitive, high speed alternative technologies such as GaAs can be used.

New interface architectures are also being used to improve the bandwidth of the RAMs. The most common is the synchronous interface. Making a memory synchronous puts it under the control of the system clock. "Wait states", during which the processor must wait for the output data from the RAM, can be reduced or eliminated. With a synchronous RAM the input addresses can be latched into the memory, freeing the processor to perform other tasks until the data is available after a known number of cycles. While the intrinsic speed of the RAM does not increase by the addition of a synchronous interface, the effective speed in the system increases since the processor no longer waits for the RAM to operate.

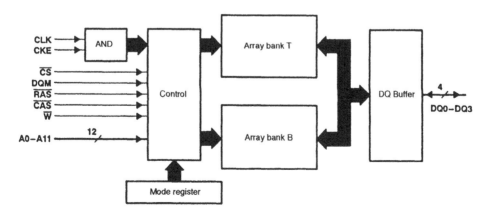

Figure 1.4 Block diagram of a two-bank synchronous DRAM (source: Texas Instruments)

Chapter 3 explores techniques for fast SRAMs further, and Chapter 5 explores techniques for fast asynchronous DRAMs.

1.3   Memories in Computer Systems

Different types of memories are used in the various parts of a computer system. A block diagram of a typical computer system is shown in Figure 1.5. The system consists of the cache subsystem, the main memory, the graphics subsystem and the peripheral devices such as printers.

The cache subsystem is a result of the discrepancy in speed capability and price between the SRAMs and DRAMs. This led to a split of the main memory into a hierarchy in which a small fast SRAM cache is inserted in the system between the microprocessor and a larger bank of slower but less expensive DRAM main memory. The cache holds data which has a high probability of being wanted next by the processor, so less time is spent accessing the slow banks of DRAMs. If the cache has the data which is wanted, it is called a "hit"; if not, it is called a "miss". Many of the attempted solutions to the bandwidth mismatch problems in computers are aimed at improving the performance of the main memory cache hierarchy, or making it less expensive.

The main memory is the active storage area of the computer. It is normally multiple megabytes in size. To reduce the cost of the system the main memory in all but the smallest computers is composed of DRAMs. These DRAMs also need to be fast since the average speed of the entire system is reduced if a cache miss takes too long to service. Since a large share of the main memory DRAMs go into the PC market, it is also necessary that the DRAM modules be fast since modules have become the standard component of the PC.

The large proportion of DRAMs that are shipped to the desktop, laptop, and graphics subsystem part of the computer market is shown in Figure 1.6.

The tendency to reduce the cost of the desktop PC by shipping it with less than optimal memory makes the add-on DRAM market of significant size. Most of the add-on market is DRAM modules for desktops and portables.

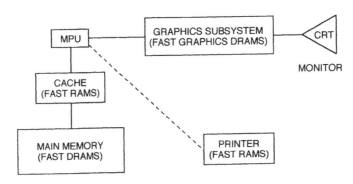

Figure 1.5   Block diagram of typical computer system

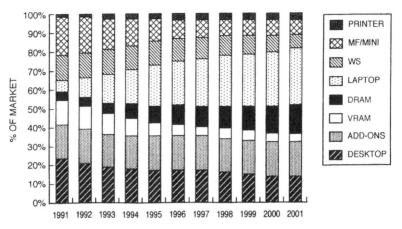

Figure 1.6   Estimated DRAM shipments to computer market by segment (%) (source: Memory Strategies International)

Graphics subsystems have special requirements. They require a fast random port to communicate with the processor and a fast serial port to update the display CRT. These requirements can be met in the graphics subsystem in the logic or the memory, or in a combination of the two. Figure 1.6 shows the trend toward fast single port graphics DRAMs and away from multi-ported graphics VRAMs.

The next three sections give an overview of these three computer divisions and the memories optimized for them.

1.3.1   Cache SRAMs

Microprocessors with 1–2KB embedded SRAM caches began to appear in about 1987. Most microprocessors today contain fairly sophisticated embedded first level (L1) SRAM data and instruction caches. These L1 caches tend to be partially to fully associative and achieve high bandwidth using the wide bus architectures possible with embedded memories.

Since the maximum size of memory that can be cost-effectively embedded in a logic chip is small, many systems from workstations to PCs also use an external second level (L2) fast SRAM cache.

The required size of this L2 SRAM cache tends to grow as the system demands on it increase due to more sophisticated operating systems and applications. Figure 1.7 indicates an estimate of fast SRAM shipment by computer equipment segment. The rapid growth of the SRAMs shipped for cache applications in portable and desktop PCs can be seen.

Special architectures are frequently used in the SRAM caches since the effective speed of the processor can be increased many fold if the SRAM cache hit rate can be improved through architectural changes in the SRAM or in the cache subsystem logic or both.

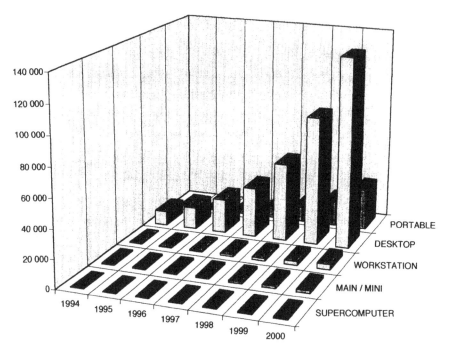

Figure 1.7 Fast SRAM shipments by equipment segment (thousand MBytes) (source: Memory Strategies International)

In the simplest cache, called "direct mapped", each address corresponds to an address in main memory and the hit rate is proportional to the memory density. A larger cache has a higher hit rate. The trade-off is that while the cache controller for a direct mapped cache is simple and available off the shelf at low cost, the cost of the SRAM for the larger cache is high.

An alternative is to use a smaller SRAM with some level of associativity. An associative cache is one in which the same bit of data can be stored at multiple locations in the cache. In a fully associative cache, data can be stored at any location, but the architectural complexity makes it very expensive.

Much of the benefit in increased hit rate of an associative L2 cache can be obtained, however, from a two- or four-way partitioning of the memory which is considerably less complex. In a two-way set associative cache, for example, the data can be stored at two possible locations in the cache.

Modern processors have included the capability of using data in high speed bursts. This has led to burst access modes being implemented on the fast cache SRAMs.

SRAMs are also used for cache buffers between multiple processors in a system or between the processor and its peripherals.

When multiple systems attempt to access the SRAM, dual port capability is frequently added to the SRAMs to gain the bandwidth required by the system. SRAMs configured both with random access and as FIFOs have been developed for this type of application.

Chapter 4 explores the topic of cache SRAMs further.

1.3.2 DRAMs in High Performance Main Memory

The large size of the main memory of a PC or workstation has dictated that it be composed of low cost DRAMs. The main memory is, however, merely the next level to the SRAM cache. If a miss occurs on the cache, the processor must access the main memory. If the DRAMs are too slow, the average access time for the system can drop significantly. The DRAMs must offer high speed random access as well as also offering high speed burst access.

The DRAMs have met the speed challenge by using alternative output modes to access fast bursts of data from the wide internal "page" of the RAMs. All of the data from a single page of the DRAM can be accessed randomly. Older access modes such as static column and nibble mode have given way to an evolution of page mode from fast page to hyperpage (EDO) to burst EDO (Extended Data Out).

In a DRAM read operation, all of the data on the selected row (page) in a DRAM array appears on the sense amplifiers when that row is selected. In random access mode, only one column address is accessed. In page mode, the data from the selected row is held on the sense amplifiers while multiple column addresses are selected from it. This eliminates the time required to write the information from the sense amplifiers back into the memory cells and perform precharge before another row address access can take place. In a 16M DRAM with 60 ns random access time, the page mode access time is typically 30 ns.

Hyperpage (EDO) mode changes the output control so the data remains on the bus for a longer time. It can increase the effective cycle time by as much as 30%.

The interface of the DRAM has been made synchronous in most high speed DRAMs including the JEDEC SDRAM I and SDRAM II (DDR), the SLDRAM, Rambus DRAM, CDRAM, ESDRAM, VC-DRAM, and MDRAM. Both inputs and outputs are synchronous in most cases.

The width of the DRAM has also increased from the 1-bit interfaces of the 1980s to as wide as 16 bits for main memory. In addition, DRAMs in PCs normally use DRAM SIMMs (Single-In-line Memory Modules) for main memory. SIMMs are small PC boards containing multiple RAMs. These boards are used like individual components which fit vertically into the main PC board. They increase the memory width up to that of the system bus, ranging from 32 bits in a small PC up to 200 bits in a workstation. Double-sided SIMMs, called DIMMs, are also widely available.

Memory bandwidth can be doubled by doubling the width of the memory bus on the SIMM. If the memory bus width stays the same, this doubles the amount of system memory. It is, therefore, most useful in larger systems which require the additional memory in any case. If, however, the DRAM width can be doubled while halving its depth, then the wider bus does not change the amount of memory in the system. The issue here is granularity, which is the minimum memory expansion increment of a system. The wider a memory of a given density, the lower is its granularity.

An additional speed advantage sported by the JEDEC Standard Synchronous DRAM and the Rambus DRAM is multiple banks on a single RAM. The multiple banks permit faster random access by permitting one bank to precharge or be refreshed while the other bank is being accessed. Multiple rows on these parts can be simultaneously open and accesses can be interleaved on the chip between the banks.

Multiple internal banks also help small fast systems with the memory granularity problem. Additional speed can be achieved by interleaving the banks on one chip rather than by interleaving multiple banks in the system which can add the cost of unneeded memory.

Another possibility in a small (sub)system is to combine the SRAM and the DRAM in one chip. High bandwidth transfers can occur between the two due to wider internal buses of both memories. Examples are the video DRAMs and field memories, the various cache DRAMs, and the FIFOs. The drawback with adding circuitry to the standard DRAM is that the cost goes up as the chip and package size increases and the DRAM testing becomes more expensive.

Both the Synchronous DRAMs and the Synchronous SRAMs feature burst mode accesses which are compatible with most modern processors. These are very fast accesses of small amounts of data following an initial access at a beginning memory address at the normal speed of the memory. The addresses of the subsequent bits of data in the burst are generated automatically by the RAM.

These burst mode accesses, like page mode, take advantage of the fact that the internal bus of the RAM is wider than the external bus. This permits all of the data from a series of burst mode addresses to be fetched from the RAM databank to its outputs upon the entry of the first address. This data can then be fed out of the RAM at the speed of the output switching transistors.

Another advantage of synchronous memories is that the system clock edge is the only timing strobe that must be provided by the system to the memory. This reduces the need to propagate multiple timing strobes around the PC board or module.

Chapter 6 explores synchronous DRAMs further.

1.3.3   DRAMs in Graphics Subsystems

There have also been attempts to use alternative architecture approaches to match the RAM output to the application so that most effective performance is achieved.

The Synchronous Graphics RAM is a fast JEDEC standard SDRAM that has added various graphics-oriented logic features such as write-per-bit and block write. It is expected to be used in low end unified memory and in midrange PC graphics subsystems where its 32 bit width provides high bandwidth and low granularity.

For higher end workstation graphics, Video DRAMs have added a serial port. The random port interfaces with the processor or controller and the serial port feeds very fast data to the video display by means of a wide parallel transfer internal to the RAM between the array and the serial register.

The serial register increases the cost of the VRAM, however, and the larger, more expensive package size also requires more space on the PC board. The testing procedure is also more complex for the VDRAM which increases the manufacturing cost of the memory.

Another video approach is the Frame Buffer DRAM which is used in the display circuitry of televisions. It has both serial inputs and outputs so it does not need the many address pins required by random access memory. A smaller package can be used due to the reduced number of pins so that, in theory, the frame memory can cost less than a standard DRAM. The frame buffer DRAM is primarily limited to this particular application.

An alternative application for wide DRAMs is the printer or fax machine which requires a very wide, low granularity memory for the line buffer.

Chapter 7 discusses graphics DRAMs in more depth.

1.4   Effect of Electrical System Characteristics on Speed

This brings us to the electrical characteristics of the input and output gates of the RAMs. These external interface characteristics must be standardized to be compatible in the system with components from a wide range of vendors.

3.3 V LVTTL gives lower power dissipation than 5 V TTL; however, to gain speed at 3.3 V, a lower swing interface is needed. Several have been standardized including: GTL, CTT, HSTL, and SSTL by JEDEC; and SLIO for the SL-DRAM by the IEEE 1596.4 committee. The DRAM from Rambus, Inc. also uses a proprietary low swing interface. All of these interfaces control for transmission line characteristics of the system by means of a reference voltage level supplied either on the RAM or externally. The design of the output buffers also must take into account whether the RAM is operating in a point-to-point environment or on a parallel bus, and whether the bus is terminated or unterminated.

Clock doubling effectively gains speed by accessing the memory on both the rising and falling edge of the clock. This requires more control of the clock but appears to be necessary to attain speeds over 200 MHz. A differential clock technique is used in both the Rambus and SL-DRAM schemes to reach speeds of 600 MHz and beyond.

Parts are beginning to be designed for these different system environments such as a synchronous ×36 organized BiCMOS SRAM from Motorola which has GTL type output buffer options for point-to-point, and for parallel terminated and unterminated buses.

Chapter 8 looks further into the questions of high speed interfaces and fast memory test.

1.5 Effect of Packaging on Speed

Moving outside the memory chip itself, the pinout and package also can affect the speed of the part. New pinouts are being used for high speed synchronous SRAMs and DRAMs which minimize the self-inductance of the package leadframe by increasing the number and placement of power and ground pins and thereby reduce the ground bounce which slows down the device when wide outputs switch.

Smaller packages with reduced lead inductance are also being used to improve speed, such as the TSOPII, the TQFP and the BGA.

Memory modules can improve speed both by moving the components closer together thereby shortening the wires between them, and by structuring the transmission line characteristics of the RAMs on the modules. These modules range from

SIMMs and DIMMs with packaged components to multichip modules with mounted bare die.

Further discussion on packaging of high speed memories and the effects of packaging on speed can be found in Chapter 9.

Chapter 2 will consider further the systems requirements for the high speed memories.

Bibliography

1. Prince, B., Speeding up the system memory, *IEEE Spectrum*, February 1994, 38–41.
2. Prince, B., *Semiconductor Memories*, John Wiley and Sons, 1991.
3. Comerford, R. and Watson, G.F., Special report on high speed DRAMs, *IEEE Spectrum*, October 1992, 34.
4. Quinnell, R. A. High speed bus interfaces, *EDN*, 30 September 1993.
5. MOS Memory Data book, Texas Instruments, 1995.
6. *SLDRAM Datasheet.*
7. *Direct Rambus Datasheet.*
8. EIA JEDEC Standard JESD8, (www.eia.org).

2 High Performance Memory Applications

2.1 The Concept of a High Performance Memory

Fast systems are a moving target. What was fast yesterday may not be considered fast today. Also what is fast for one type of system may not be fast for another. For example, a fast calculator is slower than a slow workstation. Speed also differs within the system. Not all buses run at the same speed. A 175 MHz DEC Alpha chip or a 100 MHz 486DX4 chip run externally much slower.

There are also different speed capabilities for the different memory types. Considering RAMs only, SRAMs, for example, tend to be faster than DRAMS in a comparable technology. DRAMs, in turn, are faster than hard drives, which tend to be faster than optical drives.

Most memory databooks claim that all of their parts are "fast", even the slowest ones, and by some standard they probably are fast.

The data access time is the most basic speed associated with the semiconductor memory. It depends on the memory cell type, on the technology, on the logic used in the periphery of the memory for decoding, sensing, and driving, and on the interface type and logic.

In this book a fast memory is one that is faster than the basic speed imparted by the technology.

A high performance memory goes one step further and delivers data at a faster rate than a low performance memory. The performance of a memory can be improved in other ways than by increasing the speed of the output. Improving the datarate by widening the memory interface can be the same as increasing the speed. A 50MHz 4 bit wide RAM delivers the same amount of data in the same time as a 25 MHz 8-bit wide RAM.

Performance can be enhanced by changing the memory architecture to match that of the system without changing the speed of the interface. For example, a multi-port DRAM (VDRAM) improves the datarate (bandwidth) of the system by adding an extra serial port to refresh the CRT screen. A dualport SRAM buffer improves the interface between two processors or between a fast processor and its slower peripherals.

2.2   System Architecture Determines Performance

Performance can also be improved by structuring the system architecture.

The computer hierarchy uses a small amount of the very fast but more expensive SRAMs as "cache" to hold the information that the processor is most likely to want next. This data is exchanged using various algorithms with other data held in less costly "main memory" as the next most likely data required changes.

At the next level of the computer hierarchy, information which will not be required by the current program, such as other software or database information, can be removed from DRAM main memory and held in the slower hard or floppy drive to be retrieved when needed.

Interleaving the data out from two memory banks or modules doubles the effective data rate and speed of the memory. A 2-way set associative cache subsystem can enhance the hit rate over the simple direct mapped SRAM cache.

This chapter will discuss some of the applications requiring fast memory and the various system considerations.

2.3   Systems Applications for High Performance SRAMs

2.3.1   Overview of Fast SRAM Applications

The major application areas for fast SRAMs are shown in Table 2.1. The major applications for fast SRAMs are in cache for workstation, personal computer and embedded systems, and in main memory for supercomputers. The relative size of these computer system markets is shown in Figure 2.1.

Table 2.1   Fast SRAM applications by market segment

Segment	System	Application	Speed
Computer	Desktop	Cache	Fast
	Notebook	Cache	Fast
	Workstation	Cache	Fast
	Supercomputer	Main memory	Very fast
Communication	ATM	Fast switching	Fast
	MPEG2	Decompression storage	Fast
Industrial	Embedded systems	Cache	Fast
	Space	Radiation hardened	n/a

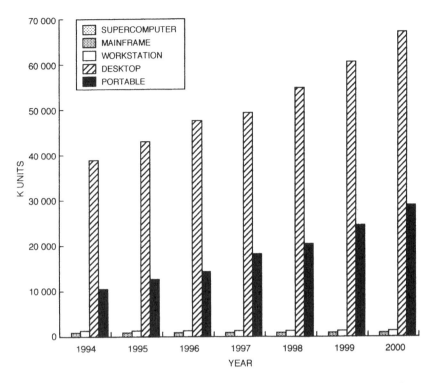

Figure 2.1 Computer end equipment markets for fast SRAMS (thousand units) (source: Memory Strategies International)

The supercomputer market is small and uses very fast specialty bipolar or BiCMOS SRAMs. The number of SRAMs used per system is large, however, with gigabytes of very fast memory in the range of 500 MHz and up required in a single system. The high speed requires very specialized cooling techniques. High speed interfaces such as ECL and specialized packaging are also required.

The largest single application for the high performance SRAMs is in fast computer cache and in fast buffers in multiprocessor systems.

While the number of systems of workstations and mainframe/servers shipped is small compared to PCs, these applications have been important since the components that are used in the mainframes and workstations tend to move downstream over time to the higher volume PCs. Also the caches of the workstations and servers tend to be multiple interleaved banks of the same caches used in the high end PCs making the workstation and server more of a factor in the SRAM markets than the number of systems shipped implies. The same SRAMs are, therefore, used in these applications and they will be discussed together.

The communications and network applications for SRAMs are primarily in emerging markets. Both the ATM switch and MEG2 compression markets are expected to be major applications in the future. Both use small quantities of very fast memory and are, hence, of interest as fast SRAM applications.

Mobile communications is a rapidly emerging market which is likely to use SRAMs. The memory requirements in small handsets are stability, random access, medium speed and some small amount of non-volatility which can be provided with batteries. As software capability from PCs moves downstream to handheld computers, it is expected that fast, easy to use, low power memories, possibly SRAMs, will be needed.

The main industrial application for fast SRAMs is as cache in embedded systems. The processors used along with the system speed and configuration tend to be similar to those in workstation applications. Since it is difficult to separate this area from normal high speed workstation applications, the cache in the embedded processors will be considered with the workstation category.

2.3.2 Systems with Fast Caches

The fast cache SRAM market is being defined primarily by the newer burst mode microprocessors such as the 586 and 686 class processors, the Power PC and Digital Alpha. The total shipments of computer systems by processor type forecast over time are shown in Figure 2.2. This includes PCs, workstations, and mainframe/servers.

The configuration and speed of the fast SRAMs that are used in these systems depend on the size of the cache and the processor which determines the speed of the system. Figure 2.3 divides the systems by cache size.

The few older 486 type systems that have cache are expected to phase it out as the faster DRAMs begin to handle the cache applications out of main memory. 586 and 686 class systems all have cache starting with the 256KB cache common in the 1995–

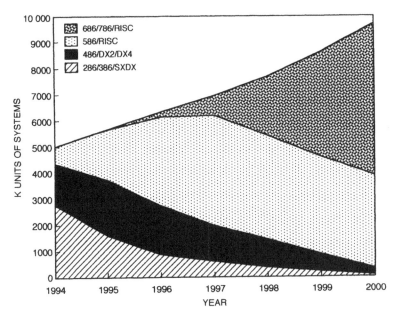

Figure 2.2 Computers shipped by processor type (thousand units) (adapted from Microprocessor Reports)

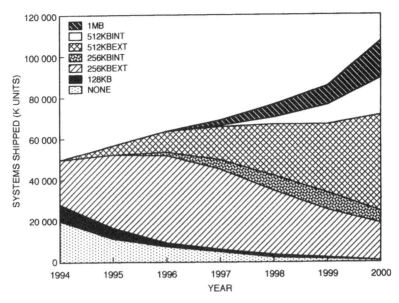

Figure 2.3   Shipments of systems with different size caches (thousand units) (source: Memory Strategies International)

1996 timeframe and moving to the 512KB and 1MB cache as newer applications require the larger memory.

Fast DRAMs are not expected to replace cache in any but the low end systems in the timeframe of this estimate for several reasons. The SRAMs will continue to be faster than the DRAMs while the speed capability of the processors already exceeds that of the fast SRAMs and continues to increase maintaining a performance advantage for the cache hierarchy architecture.

Since the cache SRAM is a small memory bank that sits close to the processor the speed loss in transmission is minimized. The trend toward smaller packages and multichip modules will also tend to get the cache SRAM even closer to the processor. A small amount of SRAM positioned close to the processor will continue to be faster than a larger amount of DRAM in the main memory bank which can not all be close to the processor.

Not only are delays in the wiring between the DRAM and the processor difficult to avoid, the wiring will also tend to be an unacceptable source of EMI radiation at very high speeds.

2.3.3   *Synchronous and Asynchronous SRAMs in Fast Caches*

As faster SRAMs have been required, the synchronous SRAM has been developed. The addition of input and output registers puts the cache under the control of the system clock for fast burst accesses. The penalty paid is the initial multiple cycle delay, or latency, at the beginning of each burst.

An analysis adapted from Micron Technology on the system benefits of the asynchronous compared to the synchronous burst SRAMs in the same technology is shown in Table 2.2.

The number sequence in the table represents the number of cycles for first through fourth access of the memory. For example a 20 ns asynchronous SRAM has 2-1-1-1 performance on a 33 MHz bus. This means that it takes two cycles to set up the first access and then one cycle for each succeeding access. The wait state in the "burst" part of the cycle is zero. For most system designers the latency in the first cycle is less important than the number of cycles to access the rest of the burst.

We assume a 15 ns asynchronous SRAM in 0.5 micron technology as a baseline. At 40 MHz the 15 ns asynchronous SRAM and the burst SSRAM both provide a two-cycle initial latency and no latency during the burst. At this frequency the asynchronous SRAM would be used.

At 50 MHz the asynchronous SRAM is 3-2-2-2, while at 50 MHz the burst SSRAM in the same technology is 2-1-1-1 with no wait state in the burst. The ability to avoid a wait state means that at 50 MHz most systems would switch to the flowthrough burst SSRAM if the price is competitive. They would not normally switch to the pipelined burst SRAM at 50 MHz since the initial latency is longer than that of the flowthrough part and both have 0 wait states during the burst at this speed.

At 75 MHz the flowthrough burst SSRAM cycle becomes 3-2-2-2, and the pipelined synchronous burst SRAM is still 3-1-1-1 with no wait state in the burst. At 75 MHz therefore many systems would find the pipelined burst SSRAM more attractive.

While some systems might have a problem with the initial three-cycle latency of the pipelined part, the alternative is a double-buffered system or a more expensive cache configuration which is prohibitively costly except in the highest end systems.

The above rationale assumes, of course, that the cost differential between the parts at the trade-off speed is essentially zero. The cost differential in volume, however, of all three parts should be similar since they are assumed to be in the same technology. Package cost might cause some price differential.

The conclusion is that a 15 ns asynchronous SRAM could give 0 wait states to about 40 MHz, while a 9 ns burst SSRAM in the same 0.5 micron technology could give 0 wait states to 66 MHz and a 7 ns pipelined burst SSRAM could give 0 wait states over about 75 MHz.

Similar reasoning leads to the conclusion that fast DRAMS are unlikely to replace SRAM cache in any but the lowest end systems due to the inherent slow speed and resulting higher latency of the DRAM [6]. For example, with a 50 MHz bus, a pipelined burst SSRAM has a burst read speed of 50 MHz with a 3 cycle latency (3-1-1-1) and a Hyperpage Mode (EDO) DRAM has a burst read speed of 25 MHz

Table 2.2 Analysis of SRAM types by bus frequency

System bus (MHz)	33	50	60	66	75	83	100
Async SRAM	2-1-1-1	3-2-2-2	3-2-2-2	3-2-2-2	3-2-2-2	3-2-2-2	3-2-2-2
Sync burst SRAM	2-1-1-1	2-1-1-1	2-1-1-1	2-1-1-1	3-2-2-2	3-2-2-2	3-2-2-2
Pipelined SB SRAM	3-1-1-1	3-1-1-1	3-1-1-1	3-1-1-1	3-1-1-1	3-1-1-1	3-1-1-1

Source: adapted from Micron Technology [2].

with a 6 cycle latency. In a 50 MHz system the EDO DRAM would be used with one cycle of latency in the burst (6-2-2-2).

2.3.4 Cache Size and Speed Requirements of Computer Systems

An analysis can be done of the various PC and workstation categories to determine the cache size and speed requirements of these applications.

The market can be split roughly into four ranges of systems; the highest is workstation, followed by high, mid-range, and low end PCs. These are normally defined by price ranges for the various systems as shown in Table 2.3.

Processor and associated cache speed choices for these systems in 1995 are as shown in Table 2.4.

In the 486 chips, a phase lock loop (PLL) on the processor doubles the on-chip speed. For example, the 486 is a 33 MHz chip and a DX2 doubles that to 66 MHz on chip and the DX4 doubles it again. The 486 has a 32 bit bus and runs generally at 5 V.

The 586 class processor's 64-bit bus is clocked at 66 MHz to work with the fast cache SRAMs such as the Burst SSRAMs. A 9–10 ns Burst SSRAM is required to support a 66 MHz Pentium rather than a 15 ns SRAM since there is a 5 ns address set-up time. The newer Burst EDO DRAMs can also support a 66 MHz bus speed. For the faster internal clock speeds up to 120–150 MHz the Pentium must also use an internal clock doubler.

Table 2.3 Price ranges defining classes of computer systems

Class	Price range
High end PC	$1500–2500
Mid-range PC	$1000–1500
Low end PC	$500–1000

Table 2.4 Processor speed choices

Processor	Volts (V)	Speed (MHz) int.	ext.	Bus Width (bits)	Speed of ext. cache (ns)
486DX2	5.0	66	33	32	15
486DX4	5.0	99	33	32	15
586	3.3	133	66[a]	64	9
586	3.3	120	66[a]	64	9
586	3.3	75	66[a]	64	9
586	3.3	66	66	64	9
P6	2.9	120	–	64	Internal

[a]Some Pentium systems are clocked externally slower or at half speed (33MHz).

The initial 686 class processor, the P6 from Intel, is also designed with an external bus that operates at half, a third or a quarter of the CPU clock speed. This is not used with cache SRAM in the initial P6 offering since the cache chip is integrated into the memory package and is on a private bus. If, however, later high volume commodity P6s are sold without the integrated cache, then the speed doubling could be used to support a slower cache in much the same way it is done with the 486 class of processor.

Doubling the speed on the chip was done in part to ease the bandwidth requirements on the memory. However, it was also done to reduce the transmission effects in the wires between the processor and the memory in the system. Since 66 MHz on the motherboard would not pass FCC regulations for EMI, but it would in the smaller dimensions on the chip. It was, therefore, doubled on chip rather than on the board.

It is expected that as systems move to faster processors and memory, there will be a lot of effort to tighten systems with smaller packages and the use of compact module assemblies.

Another consideration is that for systems with a 32 MHz PCI specification, the processor may be clocked at a slower speed. So, for example, a 66 MHz Pentium may still be clocked externally at 33 MHz to be compatible with a PCI bus.

The various Power PC chip versions have multiple options. They have options for bus width of 32 bits or 64 bits. They have options for 5 V or 3.3 V. They have options for doubling the speed on chip.

In 1995 486 class processors were used only at the low end of the desktop PC range. The 586 class processors up to about 90 MHz were used in the mid- to high end of the PC system range. Systems using the early 120–150 MHz 586 type processors, which were just being introduced, fell into the workstation category as did early systems evaluating the first P6 prototypes. Notebook PCs generally drop a grade in system characteristics from the PCs.

These requirements together with a comparison for 1998 are shown in Table 2.5.

For the workstation application, where cache is used, the speed is achieved by using a double cache and interleaving it in the system. This is an expensive board technique that is not feasible in a lower price system. For most low end applications, the cache may be eliminated in favor of a unified memory using one of the high speed DRAMs in the main memory.

The speed capability of the SRAM classified by access time is shown in Table 2.6.

Table 2.5   Estimate of cache requirements by system type

Type	Price (k$)	1995			1998		
		Cache	MPU	(MHz)	Cache	MPU	(MHz)
Workstation	<2.5	512KB	P6/	150	1MB	686	300
High end	1.5–2.5	256KB	586	100	768KB	686	125
Mid-range	1.0–1.5	256KB	586	80	512KB	586	100
Low end	0.5–1.0	128KB	486	33	256KB	586	90

Table 2.6 SRAM speed capability

Processor	Bus	TAC (ns)	Type	Mode
486DX2/DX4	33MHz	20	Asynchronous	Flowthrough
586	66MHz	9	Burst synchronous	Flowthrough
586	75MHz	7	Burst synchronous	Pipeline
686	100MHz	4	Burst synchronous	Pipeline

The 512KB cache workstation, that was originally forecast to be the first user of the 586 class processors, was expected to run at 66 MHz on the bus and required a 7ns SRAM in the cache. Parity was also used in this system so the SRAM was required to have ×9 or ×18 organization.

A 256Kb SRAM organized 32K×9 was developed with a design target of 7 ns speed. These SRAMs proved difficult to build with the required speed and were too small for the application. Early 64K×18 synchronous SRAMs were then built for the high end 512KB workstation cache.

It then appeared that the first users of the Pentium would be PCs not workstations. To improve SRAM supply and reduce the price, a 32K×32 Burst SSRAM was developed by many suppliers with flowthrough access time of 9–10 ns but that could run in pipelined mode at 6–7 ns in the system. This was followed by 8 ns flowthrough parts that ran at 4–5ns in pipelined mode and a density increase to 64K×32. Parity was also dropped since it made the SRAM more expensive and was not needed in a typical PC. Parity was retained for parts used in workstations.

High speed interfaces, such as HSTL, GTL and CTT, have begun to appear on SRAMs. Intel combined the L2 cache and the P6 processor in a single package with a GTL type of interface for the high end systems. Compact packages such as the BGA (ball grid array) were developed to reduce package inductance.

As the industry moves to faster processors a faster speed can be provided by a combination of technology and high speed interfaces. The pipelined speed of 4 ns is required at 100 MHz and can be provided by the 0.35 micron technology SRAMs.

2.3.5 SRAM Use Based on Processor Speed

For processors of 50 MHz and above the most cost-effective solution for 586 class systems is the synchronous SRAM, since an asynchronous SRAM must be in a more expensive technology, such as BiCMOS, to have an equivalent speed with an otherwise identical synchronous SRAM. Above 66 MHz a synchronous burst pipelined part is preferred to an equivalent technology synchronous flowthrough part because of the reduced access time of the pipelined burst part at essentially the same cost.

A 0.5 micron technology 32K×32 synchronous burst pipelined SRAM can provide a 5 ns access time during the burst. The initial latency, however, is 3 cycles rather than the 2 cycles of the flowthrough part. The flowthrough access time of the same part would be about 7 ns.

In cases where non-sequential bursts dominate, a flowthrough part may be required to meet the performance goals. In this case a more expensive technology will be required and the part will cost more.

Initially the external bus speed of the 90 MHz and 150 MHz Pentiums was maintained at 50–60 MHz. Intel has indicated that a 250 MHz Pentium is planned. This will likely be a multiple of the 120 MHz version and provide the equivalent of a DX4.

If we make the assumption that low cost P6s will be sold without the integrated cache and will use a voltage doubler, then 50 and 60 MHz cache SRAMs could be also used for a 100 and 120 MHz P6DX2, and a 180 MHz P6DX3. It is just as likely, however, that SRAMs with low swing interfaces and miniature packages will appear and the external bus speed of the processor will increase.

The estimated cache synchronous SRAM characteristics by system from 1996 to 1998 are shown in Table 2.7. The SRAMs are classified by pipeline and flowthrough architectures. [3]

Table 2.7   Estimate of cache SSRAM characteristics by system, 1996–1998

	1996				1998			
System	Bus (MHz)	MPU	Tech. (μM)	SSRAM PL/FT	Bus (MHz)	MPU	Tech. (μM)	SSRAM
Workstation	133	686	0.35	PL	200	786	0.3	PL
High end PC	66	586	0.4	PL	133	686	0.35	PL
Mid-range PC	50	586	0.5	FT	66	686	0.4	FT
Low end PC	33	486	0.7	FT	50	586	0.5	FT

2.4   Overview of Applications for High Performance DRAMs

The major system applications for high performance DRAMS are shown in Table 2.8.

The estimated percentage of the DRAM market for each of these application areas from 1991 to 2001 is shown in Figure 2.4. The computer segment is split out into printer, mainframe, workstation and PC.

Application areas growing faster than the overall market include PC, Printer, and Communications. Consumer and Industrial grow with the market and Workstation and Mainframe are expected to grow less than the market.

2.5   Main Memory Applications for DRAMs

The cache hierarchy gives a general framework for the organization of computers. The SRAM cache enables the system to more fully utilize the processor's capabilities. SRAMs, however, are more expensive than DRAMs so systems tend to minimize the amount of SRAM used. DRAM main memory sits below the cache subsystem and is faster but more expensive than the archival memory. Main memory requirements differ by system type.

Table 2.8 Applications segments for fast DRAMs

Computer:
    Main memory (workstation, PC, notebook)
    Graphics subsystems (frame buffers, video)
    Add-on modules (SIMMs, DIMMs)
    Peripherals (printers, fax)

Consumer:
    Mobile systems (notepad, communicator)
    Video (TV, VCR)
    Games (multimedia)

Communications:
    Switching systems
    Set top (ATM, MPEG2)

Industrial:
    Medical (X-ray video, scanners)
    Embedded processors

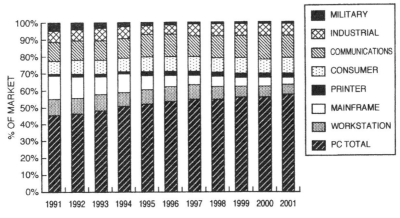

Figure 2.4 DRAM market by major applications segment (%$) (source: Memory Strategies International)

2.5.1 *Mainframe and Supercomputer Applications*

Mainframe and supercomputer applications fall into two types, those that have single, fast, large main memories and those that are multiple smaller workstations operating in a multiprocessor arrangement. These latter are much more common and will be considered as part of the discussion on workstations.

The older type has a very large fast main memory in the gigabyte density range. These main memories use very deep bit wide organizations of DRAMs.

For example a 160 MHz, 512MB system with 128-bit bus could use two interleaved boards in main memory, each board having 128 16M×1 Synchronous DRAMs (SDRAMs) with 80 MHz burst access time. The interleaving provides the speed doubling and would require also high speed connectors and logic components. The minimum granularity for this system is 512 MB, since memory must be added in 512 MB increments. The bandwidth (datarate) for this system is 12.8 bits×160 MHz =2.6 GB/sec.

Such large granularity is rare and few 16M×1 devices were actually manufactured. A computer that required this density of memory and high bandwidth would probably use 16M×4 DRAMs.

2.5.2. DRAMS in Main Memory in Workstations

Base main memory requirements for workstations range from 32 to 64MB of main memory with upgrades possible, usually with performance enhancements gained by interleaving boards.

64MB of DRAM on a 128-bit bus can be achieved using one bank of 32 4M×4 DRAMs. If we assume a 20 ns hyperpage mode (EDO) cycle time for the parts, then the system runs at 50 MHz.

If two banks are interleaved, upgrading to a 128MB system, then one interleave gives a 100 MHz speed. The use of a×1 organization would make the granularity too large, and a ×8 would make it too small and might also reduce the speed if ground bounce is taken into account.

Table 2.9 illustrates possible main memory organizations for minimum granularity and speed for a 128-bit bus using 16M DRAMs, 16M SDRAMs, and 64M SDRAMs.

The speed column assumes a 20 ns EDO access and a 12 ns SDRAM burst access for the 16M and an 8 ns SDRAM burst access for the 64M. The speed for two interleaved boards (banks) is also shown.

While a ×4 16M DRAM gives the required 64MB base granularity on a 128-bit bus, it will take a ×16 64M DRAM to give a 64MB granularity. If a 240 MHz speed is desired, then the 64MB base system for the 64M SDRAM will require interleaving two banks of ×32 64M DRAM. Faster and wider DRAMs will be needed. High speed interfaces will be required.

The ability to interleave multiple banks on the SDRAM chips themselves may allow the speed to be increased without the need to interleave larger blocks in the system or to resort to even wider buses.

Table 2.9 Memory granularity and speed for 128 bit bus system

DRAM type	No. of banks	DRAM Granularity (MB)				Speed (MHz)
		× 4	×8	×16	×32	
16M DRAM	1	64	32	16	8	50
	2	128	64	32	16	100
16M SDRAM	1	64	32	16	8	80
	2	128	64	32	16	160
64M SDRAM	1	256	128	64	32	120
	2	512	256	128	64	240

2.5.3 DRAMs in Main Memory in PCs

Main memory for PCs, including desktops, notebooks, and add-on modules, make up the largest of the DRAM applications.

The characteristics of these DRAMs are defined to a large extent by the microprocessors which determine both the bus width and the required speed of the main memory.

A typical main memory configuration for a 586 PC system with 64-bit bus in 1996 is a 32MB base memory running with a 60 MHz system bus. These systems are likely to use 16M and 64M DRAMs in hyperpage mode or 16M SDRAMs in burst mode.

Possible memory configurations for 16–64MB of memory are shown in Table 2.10. Typical speeds are shown assuming use of EDO and burst SDRAM modes. In cases where the desired configuration is too slow for the bus, a different configuration can be used or wait states for the processor can be inserted.

The system banks could be memory modules. The configurations not surprisingly correspond with standard memory modules.

The simplest base configuration for a 16MB system is made with the 2M×8 organized 16M DRAM and a 2M×32 organized 64M DRAM. For a 32MB base memory at the 64M level a 4M×16 will be the simplest base configuration. It is clear that granularity is requiring that the memories become wider than in the past.

Possible DRAMs and access modes that can be used to obtain 80–200 MHz are shown in Table 2.11.

Table 2.10 Memory configurations for a 64 bit system bus

| MPU | Bus | Density (MB) | Memory type[a] | DRAM | | | | Speed (MHz) |
				Type	Mode	Org.	No.	
586	64	16	2×[1M×64]	16M	EDO	1M×16	8	40
586	64	32	4×[1M×64]	16M	EDO	1M×16	16	160
586	64	16	2M×64	16M	EDO	2M×8	8	40
586	64	32	2×[2M×64]	16M	EDO	2M×8	16	80
586	64	32	4M×64	16M	Sync	4M×4	16	100
586	64	64	2×[4M×64]	16M	Sync	4M×4	32	200
586	64	16	2M×64	64M	EDO	2M×32	2	40
586	64	32	2×[2M×64]	64M	EDO	2M×32	4	80
586	64	32	4M×64	64M	Sync	4M×16	4	125

[a]Memory type indicates the configuration of the basic bank and the number of banks that are interleaved.

Table 2.11 DRAM access modes for 80–160MHz speeds

DRAM	Organization	Speed (MHz)	Interface
16M DRAM with EDO	×4, ×8, ×16	40	LVTTL
64M DRAM with EDO	×16, ×32	40	LVTTL
16M SDRAM	×4, ×8, ×16	100	LVTTL, SSTL
64M SDRAM	×16, ×32	125	CTT, GTL, SSTL

By interleaving banks, speeds over 100 MHz can be achieved with hyperpage mode (EDO) and well over 250 MHz can be obtained using the SDRAM.

2.5.4   DRAMs in Add-On Modules for Main Memory

The 32-bit SIMM modules for the 8–32MB main memory sizes are continuations of the 72-pin DRAM module standard organized 4M×32 (×36, or ×40). For an 8MB base memory, this can be obtained using four 1M×32 modules with one interleave containing 1M×4 DRAMs or 1M×16 with EDO for an 80 MHz speed.

For 64-bit wide systems an 8-Byte Dual SIMM (DIMM) standard is available with EDO DRAMs and SDRAMs in configurations from 16MB to 64MB. An 8-Byte 200 pin SDRAM Dual Sided SIMM (DIMM) also exists which goes to 125 MHz and above.

2.6   DRAMs in Graphics Subsystems

Frame buffers are found in both television and computer systems and DRAMs are usually used in both types of systems [1].

2.6.1   Television Displays

A display technology commonly used in television applications is called raster scanning. It is controlled by a dedicated processor called a display processing unit (DPU) or video controller. The subsystem also includes a frame buffer, a digital to analog converter (DAC) and a display monitor as shown in Figure 2.5.

A raster scan system constructs an image using X and Y coordinates to designate points on the display screen which are called pixels. The color of the image is held as bits in the Z coordinate in a bit map generated in the display processing unit. The image is displayed as a series of horizontal lines of pixels written from top to bottom on the screen. [1]

One complete image is called a frame and is buffered (held) in a frame buffer until needed on the screen. The frame buffer is usually constructed of DRAMs with each storage cell location corresponding to a pixel that will be mapped onto the screen.

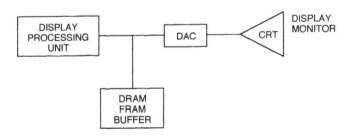

Figure 2.5   Block diagram of a simple television system

The Video Controller (DPU) scans the data stored in the frame buffer and uses it to refresh the image on the screen. The scan is from left to right and from top to bottom on the screen. Frame sizes are determined by the size and aspect ratio of the screen and are usually stated in powers of 2.

The required access time of the DRAM is determined by how fast the pixels need to be fed to the screen, called the pixel time, and is the active line time divided by the number of visible pixels per line. This is turn is determined by the refresh rate, the vertical and horizontal retrace time, and the screen resolution.

2.6.2 DRAMs in Television Related Applications

Various frame buffer types of applications in televisions require a large memory with a serial interface which can be used in series with the display for temporary storage of a frame of memory.

One of the earliest frame buffer applications was in Europe where the 50 Hz television signal caused a visible flicker which is difficult on the observer's eye. By temporarily holding a display line in a buffer and then duplicating it on the screen, the effect of a 100 Hz signal could be made. This type of television enhancement is called 100 MHz IDTV in Europe. More modern applications require a memory which can hold an entire frame and handle various minor graphics effects such as block access.

The minimum requirements for a frame memory are low cost and serial interface. The low cost requirement is met by the standard frame memory by a smaller package and chip than the normal DRAM. This is achieved by being able to reduce the number of address pins due to the serial interface, and hence reduce the size of the package. For ease of use, the frame memories have been made with pseudostatic refresh since the 1M.

More recent applications for the frame memory have included various types of stop point capabilities, and additional logic for other consumer graphics applications.

It is generally believed that high definition television systems by about 1999 will require over 30Mb (megabits) of memory. This memory needs to be low cost and specifically tailored to the application. The application can be filled with a 16M or 64M DRAM technology. It is doubtful if it will be a standard part because of the high degree of competition among the large consumer companies which has prevented the development of a true television memory standard to date. It is likely that this will remain an area for customer specific DRAM development as it has been in the past.

New TV applications such as interactive television and video conferencing which link computers and television will be covered in the next section on Communications.

2.6.3 Graphics DRAMS in Computer Graphics Subsystems

A simple graphics subsystem includes the graphics processor, the frame buffer with both parallel and serial ports, the digital to analog converter used to produce the signals required by the CRT, and the screen. The parallel and serial ports can either be produced by a logic chip with a standard single random port DRAM, or the two ports can be present on the DRAM as illustrated in Figure 2.6(a) and (b). [5]

Graphics memory subsystems in PCs and workstations must process information from the main processor and provide fast serial refresh to the display. Historically standard DRAMs have had difficulty in supplying the high datarates that are required. With the demands of new graphics software, higher frame refresh rates, increasing pixel resolutions, and added color bits, the bandwidth requirements on the DRAM are increasing significantly as shown in Figure 2.7.

To meet the bandwidth requirements, multi-port DRAMs have been used which have one random port for communicating with the processor and one serial port for updating the display. These multi-port RAMs, called Video RAMs, typically cost from 1.5 to 2 times the price of a single port RAM of the same memory density. The high end PC and low end workstation applications which have, historically, used VRAMs are increasingly coming under competitive price pressures. The faster single port DRAMs are being increasingly used by this group.

For graphics in low to mid-range PCs, it has become common to use either a custom ASIC or a commercial graphics controller chip that provides the serial and parallel ports and other graphics features, using only fast, single port, commodity DRAMs for the memory storage. The increasing datarate demands for this group of computers is putting performance pressures on the slow but cost-effective DRAM, pressures which can be met by faster single port DRAMs.

A straightforward but fairly complex calculation can be used to figure out how much single port memory is needed and how fast it will need to run given the screen resolution, number of color bits, and bus width of the graphics subsystem. If the

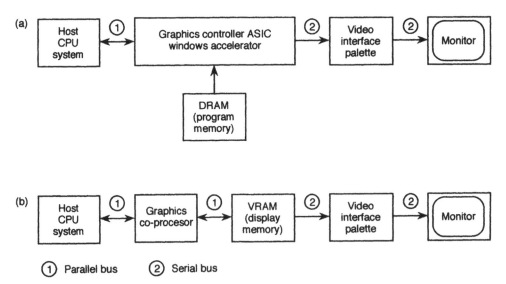

Figure 2.6   Two types of graphics subsystems: (a) with integrated GUI/RAMDAC and single port DRAM; (b) with separate GUI and RAMDAC and multi-port DRAM (source: Texas Instruments)

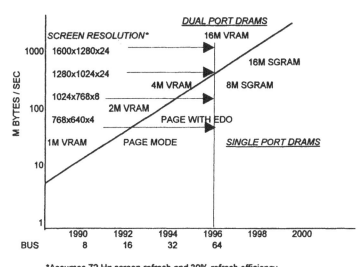

*Assumes 72 Hz screen refresh and 30% refresh efficiency

Figure 2.7 Datarate trends for graphics subsystems and graphics DRAMS

bandwidth requirements for a particular graphics subsystem are too high for a single port DRAM to support, one can choose to use a more complex memory such as a multi port DRAM.

Examples of the results of this calculation are shown in Figure 2.8 which is a plot of the data transfer rate by screen resolution and color for a 32 bit wide graphics bus assuming a 30 percent refresh efficiency. [3]

Screen resolutions up through the super VGA (1024 × 768 with 16-bit color) can use a fast single port asynchronous DRAM, while screen resolutions of up to 1280 × 1024 with 16 color bits can use a fast single port 16M synchronous graphics DRAM.

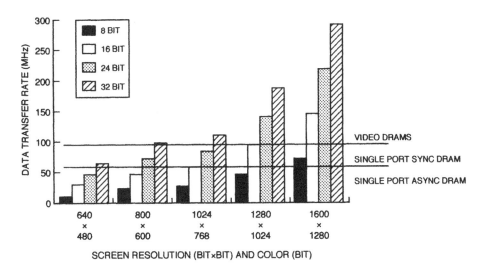

Figure 2.8 Data transfer rate by screen resolution and color for 32 bit system

Higher resolution screens, such as are found on graphics workstations, will need either multi port DRAMs (VRAMs) or other very high bandwidth applications specific DRAMs.

The method for calculation used is shown in Figure 2.9 for various systems assuming 72 Hz screen refresh. Column 1 shows the screen resolution and column 2 the number of color bits. The resulting number of megabytes of memory required is shown in column 3. Column 4 shows the bandwidth which is the number of megabytes of memory required per second.

Since a single port DRAM must both refresh the screen and support the graphics processing, only a percentage of the bandwidth is available for refresh. A common assumption for refresh efficiency in a system using single port DRAMs is 30 percent. The bandwidth required for a 30 percent refresh efficiency is shown in column 5.

This assumption more than triples the bandwidth requirement. If the width of the graphics bus used in the system is now factored in, columns 6–10 show how fast the bus must run to support the system.

For example, a 1024 × 768 screen resolution with eight bits of color on a 32-bit bus must run at 47 MHz and requires a minimum of 0.79MB of memory. This requirement could be filled by a single 8M SGRAM (1MB) in a ×32 configuration.

Graphics System Memory Requirement									
			Data Rate @ 72 Hz		Bus Speed by Bus Width (MHz)				
Resolution (Bit × Bit)	Color (Bit)	Frame (MB)	(MB/S)	(30% EFF)	×8	×16	×32	×64	×128
640 × 480	8	0.31	22.1	74	74	37	19	9	5
	16	0.61	44.2	148	148	74	37	19	9
	24	0.92	66.3	221	221	110	55	28	14
	32	1.2	88.4	296	296	148	74	37	19
800 × 600	8	0.48	34.6	115	115	58	29	14	7
	16	0.96	69.1	230	230	115	58	29	14
	24	1.44	103.7	346	346	173	87	43	22
	32	1.92	138.2	460	460	230	115	58	29
1024 × 768	8	0.79	56.6	189	189	95	47	24	12
	16	1.57	113.2	378	378	189	95	47	24
	24	2.36	169.8	566	566	283	141	71	35
	32	3.14	226.4	756	756	378	189	95	47
1280 × 1024	8	1.31	94.4	315	315	158	79	39	20
	16	2.62	188.8	630	630	315	158	79	39
	24	3.93	283.2	943	943	471	235	118	59
	32	5.24	377.6	1260	1260	630	315	158	79
1600 × 1280	8	2.05	147.5	492	492	246	123	62	31
	16	4.1	294.9	983	983	492	246	123	62
	24	6.14	442.4	1475	1475	737	369	134	92
	32	8.19	589.8	1966	1966	983	492	246	123

Figure 2.9 Technique for determining DRAM type for graphics subsystems

Screen resolutions up through the super VGA (1024 × 768 with 16-bit color) can use a fast single port asynchronous DRAM. For example, four 256K×16 DRAMs on a 64-bit bus provide 2MB of memory (satisfying the required 1.57MB). Using hyper-page mode (EDO) at 40 MHz, they provide a data rate of 320MB/sec which gives about a 35 percent refresh efficiency for a system with a 72 Hz screen refresh rate.

Some examples of potential datarates and minimum memory size using a 256K×16 DRAM are shown in Table 2.12.

It is clear that it is possible to get relatively high datarates with the 256K×16 DRAMs with EDO while retaining reasonable minimum memory size. Even if the datarate can not be sustained in an actual graphics subsystem because of the need to refresh the screen the attainable datarate is still adequate for many systems.

As an example, consider the datarate required by a screen with 1280 × 1024 × 24 resolution = 3.93MB. At a 72 Hz refresh rate with a 50 percent refresh efficiency, the requirement is 576MB/sec which can be supplied with a 64-bit bus and 4MB of 25 ns 256K×16 EDO DRAMs with either a two-way interleave or with a 128-bit bus. At a 30 percent refresh efficiency, the requirement is 943MB/sec which is difficult to meet on a 64-bit bus with asynchronous DRAMs.

Faster and wider synchronous DRAMs with graphics features can provide more bandwidth. For example, a super VGA screen resolution also could be supported by two 8M Synchronous Graphics RAMs (SGRAM) with ×32 organization run on a 64-bit bus at 80 MHz. This combination would provide 2MB (1.57MB is required) of memory and 640MB/sec of bandwidth.

Higher density SGRAMs can support higher levels of pixel resolution. A 16M synchronous graphics DRAM organized 512K×32 running at 80 MHz on a 64-bit bus could support a resolution of 1280 × 1024 with 16-bit color with at least 30 percent refresh efficiency. It provides 4MB of memory compared with the 2.62MB required.

Very high resolution screens, such as are found in workstations performing complex graphics, will need either multi port DRAMs (VRAMs), interleaved single port DRAMs, or higher speed single port DRAMs.

The fast single port DRAMs range from the standard 256k×16 and 1M×16 DRAMs with fast page mode or hyperpage mode (EDO) to 8M and 16M synchronous

Table 2.12   Datarates and minimum memory size for the 256K×16 DRAM

Bus (Bits)	Cycle time (MHz)	Interleave	Datarate (MB/sec)	Granularity (MB)	No.
16	20 page	1	40	$\frac{1}{2}$	2
16	20 page	2	80	1	4
16	40 EDO	1	40	$\frac{1}{2}$	2
16	40 EDO	2	160	1	4
32	40 EDO	1	160	1	4
32	40 EDO	2	320	2	8
64	40 EDO	1	320	2	8
64	40 EDO	2	640	4	16
128	40 EDO	1	1280	4	16

DRAMs and Synchronous Graphics RAMs. Many of them have graphics features such as write-per-bit, and block write. There are also specialty single port graphics memories appearing on the market such as the DRAMs from Mosys and Rambus.

2.6.4   *Frame Buffer Operations to Improve System Bandwidth*

There are operations that can be used on frame buffers to improve the bandwidth. Some of these are included in the DRAMs intended for graphics applications.

The Bit Block Transfer or BITBLT is included on some graphics DRAMs. It permits operation on rectangular blocks of the same height and width in a frame. The BITBLT is not restricted to contiguous blocks, operates at any pixel boundary and is able to perform a logical operation on a block that is being transferred.

Tiling is another operation that can be built into the architecture of the graphics DRAM. In tiling all of the frame is divided into blocks of a given size. All of the pixels in a tile can be operated on, permitting, for example, rapid operations to clear the screen. The tile is the maximum simultaneous area being operated on by the graphics processor. The larger the tile, the higher the pixel performance.

Double buffering is a system technique. In this case the frame buffer is divided into two separate arrays, one of which is used to refresh the display while the other is being updated. When the image on the screen needs to be changed the two buffers can be switched and the updated array is fed onto the screen. This technique improves the datarate slightly and smoothes the image transition. The drawback is that, since each of the arrays needs to be the size of the display, either twice the amount of memory is needed or the color resolution needs to be cut in half.

2.7.   Peripheral Applications for DRAMS

2.7.1   *Printers*

Printers use small DRAM buffers to hold one, or several, pages in memory during printing. Since the granularity of this application is low, the DRAMs used have been wide organizations. At the 16M density this has been a major application for the ×16 DRAM and there are even ×32 DRAMs targeted at this application. [5]

This is a high bandwidth application because of the amount of data required to feed the print engine. It has not in the past required very high speed DRAMs although the increasing capabilities of the printers may change that situation.

Some of the operations required of printers are text processing, graphics processing and numeric processing used in 3D rotations. A block diagram of a laser printer controller subsystem is shown in Figure 2.10. The width of the DRAM is required to feed the print engine.

It would be possible for a multi port DRAM to fill the function of both the DRAM and the FIFO shown. This is usually not done because there has not been a high speed requirement on the random port to justify the added cost of the VDRAM.

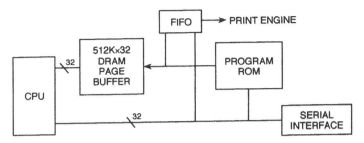

Figure 2.10   Block diagram of a laser printer controller subsystem

2.8   Consumer Applications for Fast DRAMs

Consumer applications for DRAMs include notepad computers and various games.

2.8.1   Fast DRAMs in Consumer Games

The consumer market historically has used only the slow commodity SRAMs in battery operated systems. Recently, however, the introduction of high end multi-media games has made this an application for various high speed synchronous DRAMs. The requirements are similar to those already addressed for graphics subsystems.

2.9   Communications Applications for DRAMs

Communications applications are merging with computer applications at a fast rate. This causes some overlap in the considerations of the two types of applications. In addition, some of the emerging communications applications, such as MPEG2 decompression, will be used in set top boxes which are a consumer product.

2.9.1   Digital Switching Systems

The single largest market in the communications area for DRAMs both today and for the next five years is digital switching systems. These range from massive central switching exchanges which use as many DRAMs as a supercomputer, to small office oriented PABX systems.

Telecommunications has not been an area where rapid change occurs. Systems, once put in place, are expected to last from 10 to 25 years.

The memory system characteristics of typical central switching and local switching systems in 1995 are shown in Table 2.13.

Table 2.13 Typical memory system characteristics of switching systems

	Central switching	Local switching
Clock rate (MHz)	33–50	66–100
Memory size (MB)	512–1024	64–128
Error correction	Yes	Yes
Bus width (bits)	64–128	32–64
Cache	Yes	Yes
Granularity (MB)	16+	16

These are large, fast systems which expect to use the highest density and fastest DRAM option available. It is expected that this area will use both the DRAM with EDO and the SDRAM.

2.10   Emerging Communications Applications

There are also many emerging applications in the communications area. Some of these will be considered in the next sections.

2.10.1   *Video Conferencing and Interactive TV Equipment*

One emerging application for DRAMs is that of video conferencing equipment. Both DRAMs and VRAMs should be used in this equipment. The DRAMs will be used to support the microprocessor control and the VRAMs, or other fast graphics RAMs, will be used for frame storage, frame freeze, zoom, and other features already defined in the high definition television area. A related area is interactive television.

2.10.2   *ATM Switches*

ATM switching is part of ISDN and is the digital link between the computer and communications networks of the future. The Asynchronous Transfer Mode (ATM) protocol standard uses standard length transmission packets to switch a large variety of data, voice, and video information through the same network. It is believed that the ATM WAN (wide area network) and LAN (local area network) of the future will replace conventional communications networks.

The standards issues are still not settled for such items as hooking conventional networks like Ethernet and Token Ring into an ATM infrastructure. A LAN emulation standard is still in process.

These switches need a large amount of memory which is able to operate at high speed. Some of the manufacturers working on these switches have considered using the SDRAM in the ATM switch. This is an emerging market and the memory type required is still not well defined.

2.10.3 Digital Compression

The digital compression market is also part of the computer–communications link of the future. The ISO MPEG II algorithm standard has recently emerged for high speed compression and decompression of motion picture data transmissions.

MPEG II permits a resolution of 352 × 240 pixels at a data compression rate of 1.5Mbits per second. This is sufficient resolution for VHS quality display from CD ROMS. It is expected that video compression will permit the television industry to expand services significantly.

While this is also an emerging market, it is expected that the majority of the memory applications will be in the decompression area and used in consumer set top boxes to decompress transmissions from satellite and cable stations. Television beamed from satellite using MPEG began in the US in the summer of 1994 with a few thousand subscribers.

Set top boxes are expected to find a major market also in countries that do not have an extensive conventional television infrastructure such as those in South America.

An illustration from Matsushita of a multimedia terminal on an ISDN network using an MPEG II type decompression algorithm is shown in Figure 2.11. This system is envisioned to use both a DRAM as main memory for the MPU and a VRAM for the decompression algorithm.

2.10.4 Mobile Communications

Of the four emerging areas considered, mobile communications is the largest and most rapidly growing. While most of the mobile phones today include SRAM or Flash as memory, the PDA type mobile phone of the future is likely to require enough memory to include DRAMs.

These DRAMs will be required to operate off a battery and will therefore need to have both low standby power and low active power requirements. They will also be

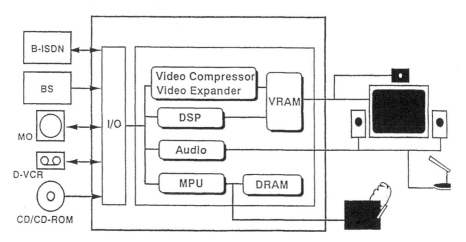

Figure 2.11   Illustration of a multimedia terminal using MPEG II (source: Matsushita)

required to run fairly sophisticated software from simple word processing programs to spreadsheets. They will also need to handle various multimedia and communications functions. For these reasons a high level of bandwidth from the memory is expected.

A power down function with self refresh will also be required for DRAMs in battery powered applications. for an asynchronous part such as an EDO DRAM, this will be a "write enable with CBR refresh" function. For a synchronous SDRAM the "power down with self refresh" command can be used.

2.11   Industrial Applications for DRAMs

Traditional industrial applications have merged with what used to be considered military applications. These two are considered together in this section.

2.11.1   Medical Systems

Medical systems use DRAMs primarily for frame storage for applications such as CAT Scanners, and real time X-ray systems used in surgery. The DRAMs required are high speed graphics DRAMS. Historically some medical systems used frame memories.

2.11.2   Embedded Controllers

The embedded controller area is small in number of systems, but they require a large, high bandwidth DRAM main memory because of the amount of number crunching done. The newest VME boards are using the Pentium and various RISC processors and are similar to those on workstations discussed previously.

The VME computing boards have historically been used primarily in military types of applications such as calculating satellite orbits, or missile trajectories. They are being used today to link together to form massive parallel processors which act as low cost supercomputers. For example, a Cray Research massive parallel processing machine links together an array of DEC Alpha-based VME boards.

Bibliography

1. Frame buffer architecture, *NEC Memory Products Databook, Volume 1, DRAMs, DRAM Modules, Video RAMs*, 1993.
2. Child, J., Faster processors ignite SRAM revolution, *Computer Design*, July 1994, 97.
3. Prince, B., Report on single port DRAMs for graphics subsystems, *Memory Strategies International*, August 1994.
4. Prince, B., Report on high density cache SRAMs, *Memory Strategies International*, June 1995.
5. *Graphics Systems, Products and Applications*, Texas Instruments, 1989.
6. *Presentation on Fast SSRAMs*, Motorola, 1995.

Fast SRAMs

3

3.1 Overview of Fast SRAMs

A major attribute of the SRAM is its speed, so fast SRAMs have continued to be manufactured in a range of densities from very fast 16Kb dual port SRAMs to fast synchronous 4M burst SRAMs. This gives the SRAM some advantages over the DRAM whose drive toward cost reduction pushes the smaller densities rapidly into obsolescence.

CMOS static RAMs are stable under temperature and voltage variations and maintain their data without refresh as long as a voltage is applied. BiCMOS static RAMs are fast but have higher power dissipation, while GaAs SRAMs are very fast and radiation hardened but are low density and more expensive.

The CMOS logic used in these SRAMs has been easy to modify into the various configurations required by differing systems, from multiple ports to associative SRAMs, to cache tag SRAMs.

In speed the SRAMs have stayed close to the processor speeds but have undergone many changes to meet the required datarates of modern processors. The pin configuration has been changed to reduce ringing in the system. The interface has been made synchronous to operate better in clocked systems. New access modes have been added to improve access to the wide internal bandwidth.

This chapter will discuss the techniques by which SRAM speed is maximized. The next chapter will discuss specifically those fast SRAMs made for cache applications.

3.2 Fast SRAM Technology

Fast SRAM technology has paralleled the capabilities of the volume DRAM technologies in line width although they have fallen more than a generation behind in density. The drawn feature size and effective channel length of a series of fast CMOS SRAM processes from 1981 to 1995 are shown in Figure 3.1.

The meaning of 'fast' has been changing with time as shown in Figure 3.2. For the 16K SRAM, 50 ns was fast; for the 64K SRAM, 40 ns was fast; for the 256K it was 30 ns; for the 1M 25 ns is fast. Today a fast CMOS SRAM is generally faster than 20 ns. Maximum specified valves are indicated.

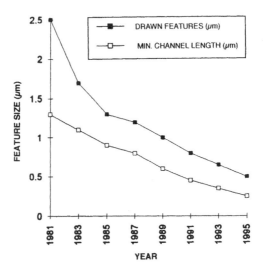

Figure 3.1   Technology by year for fast SRAMS (source: Integrated Devices Technology [10])

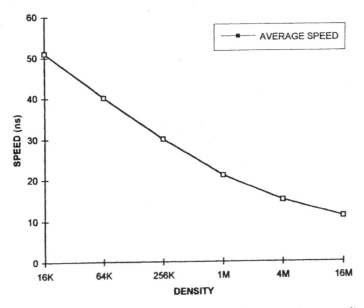

Figure 3.2   Trends in SRAM speed by density (source: various suppliers)

These speeds are associated with the first technologies that these fast parts were made in. A fast 64K SRAM in a 1.2 micron technology would be expected to be about 55 ns, and a fast 4M in a 0.5 micron technology would be about 25 ns. A fast 64K SRAM, however, in a 0.5 micron technology would be about 15 ns. The life cycle of a single density of SRAM tends to be long since it migrates to the new technologies as they are developed for the speed potential.

3.3   Architectural Influence on SRAM Speed

While a fast technology contributes to a high speed part, it is by no means the only factor. Architectural techniques are also required to allow an SRAM to achieve its maximum potential speed. Examples of architectural features which can enhance speed are separate inputs and outputs to avoid bus contention, and the use of output enable for fast data access. Architectural design features such as divided arrays and Address Transition Detection (ADT) are not within the scope of this book but are covered in reference 3.

3.3.1   *Separate and Common Inputs and Outputs*

A simple SRAM can be configured with separate or common data inputs and outputs as shown in the block diagrams of a 256K×4 SRAM in Figure 3.3(a) and (b) [1].

One of the advantages of common inputs and outputs is that a smaller package can be used because there are fewer pins. This advantage has diminished, however, with density.

For example, in the case of a 4K×4 SRAM, the common I/O part had 20 pins and the separate I/O part had 24 to handle the four additional I/O pins. This was a 20 percent increase in the pincount which factors to a 20 percent increase in package size with the same lead spacing for a 16K SRAM. At the 4M density level, however, the penalty paid for separate I/O in a 1M×4 is proportionately less, only 13 percent in going from a 32-lead common I/O package to a 36-lead separate I/O package [2].

The main advantage of separate I/O is faster read/write cycles because bus contention can be avoided. With a common I/O part there can be bus contention if data is applied to the inputs for a write cycle when data is still on the outputs from a read cycle.

This possibility for bus contention can be illustrated using the timing diagrams for a Write Enable controlled write cycle for a common I/O and separate I/O SRAM shown in Figure 3.4.

For the common I/O write cycle, shown in Figure 3.4(a), either WE\ or CS\ must be high (off) during all address transitions. The write occurs when WE\ and CS\ are both on (low). To avoid bus contention, when there is valid data on the output, the input signals should not be applied.

After WE\ goes low, the maximum write hold time, t_{WHZ}, must elapse before the data inputs can be turned on. As a result, the minimum write pulse width must be

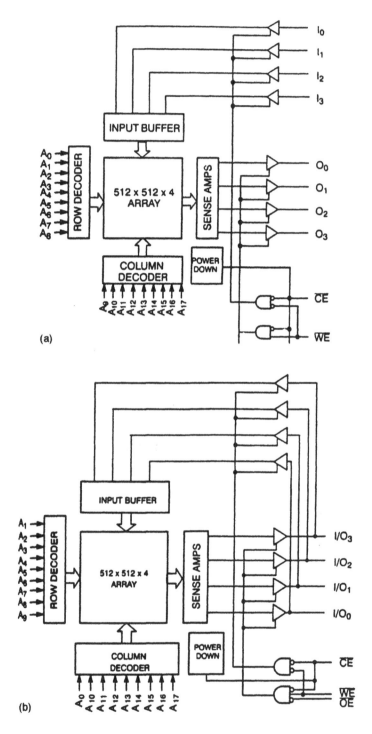

Figure 3.3   Block diagrams of 256K×4 SRAM with (a) separate and (b) common I/Q (source: Cypress Semiconductor)

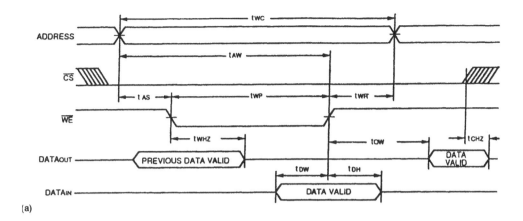

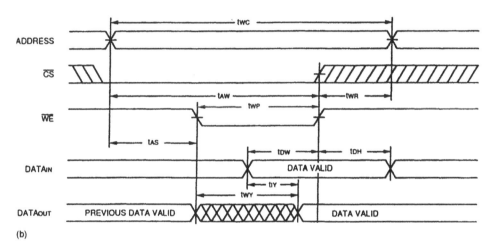

Figure 3.4 Write Enable controlled write cycle timing diagram: (a) common I/O SRAM; (b) separate I/O SRAM with outputs tracking inputs (source: IDT [10])

longer than the output turn-off time plus the time the data in must be valid before WE\ is turned off, t_{DW}:

$$t_{WP} > t_{WHZ} + t_{DW}$$

With separate inputs and outputs, the outputs either can be off (high impedance) when the inputs are on or can track the inputs. The case of outputs tracking the inputs is shown in Figure 3.4(b).

In this case, the time to turn off the previous output after WE\ turns on (t_{WHZ}) is eliminated so the timing is no longer limited by this parameter. As a result the write pulse width can be shortened and the write cycle time improved. The read cycle is not changed.

For further information on bus contention, an excellent discussion on this topic can be found in Reference 6 as well as a summary of this discussion in Reference 3.

3.3.2 Output Enable

SRAMs can also add speed in the system by the addition of an output enable control pin which turns the outputs on and off externally. Figure 3.5(a) shows a read cycle timing diagram for a 256K×4 SRAM without OE\ and Figure 3.5(b) shows a read cycle for a similar SRAM with OE\ from the same manufacturer.

The output enable provides the flexibility to have the chip addressed and selected before the outputs are turned on. The result is that data can be accessed in a much shorter time from output enable than from chip select. Typically output enable access time is around half that of address access and chip select access time. In the case of a 1M SRAM with 12 ns address access time (t_{ACC}) and 12 ns chip enable access time (t_{CO}), the output enable access time (t_{OE}) is 7 ns [2].

3.3.3 Wide Bus SRAMs for Bandwidth Improvement

The important factor in a high performance memory system is actually the rate at which data can be obtained from the memory to feed to the processor. If more data can be obtained on one access, then the datarate is faster for any given speed of access.

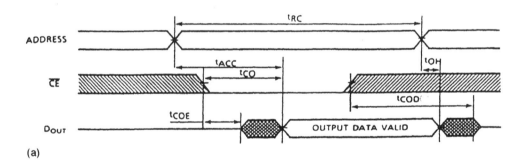

(a)

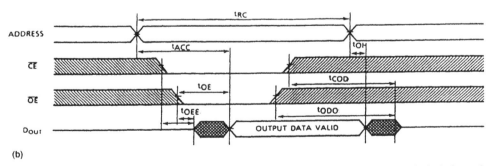

(b)

Figure 3.5 Read cycle timing of a 256K×4 SRAM: (a) without OE\; (b) with OE\ (adapted from Toshiba [2])

For this reason, the SRAMs have gone from an initial output width of 1 bit to 32 bits. While this adds I/O pins, it reduces the number of address pins. The net effect, however, is to increase the pincount. For example, a 256K×4 SRAM requires 28 pins and a 64K×16 requires 42 pins. Pinouts for the two packages are shown in Figure 3.6. The 64K×16, although needing only 42 pins, is in a 44 pin package. This is done either for potential upgrade or to fit in a standard size package, both of which are cost reduction factors.

The datarate of the two SRAMs is obtained by multiplying the width of the output by the potential speed of the output. The maximum access time specified for the 256K×4 SRAM used in the example is 12 ns (83 MHz) and for the 64K×16 SRAM is 15 ns (67 MHz). The bandwidths are therefore 41.5MB/sec and 134MB/sec respectively. Even though the 64K×16 is slower, its datarate is five times greater than that of the 256K×4. If both were used, however, on a 32 bit bus the datarate of the 256K×4 would be 332MB/sec compared to 268MB/sec for the 64K×16.

Historically SRAMs with ×1 organization were used in large fast systems to provide the depth required by the system. Since the outputs could be tied down with extra power and ground pins on the bus, a fast SRAM could give a very high bandwidth in such a system. For example, a well designed mainframe with a 1M memory depth and a 256 bit bus using 256 10 ns 1M×1 SRAMs and an adequate number of power and ground pins would run at 100 MHz with a datarate of 3.2GB/sec.

Since the depth of a ×1 SRAM is now frequently deeper than the intended system, the main configuration for large fast systems is ×4. However, there are also ×2 SRAMs made. Hitachi, for example, offers a 512K × 2 ECL SRAM for very fast deep systems such as mainframes and supercomputers [4].

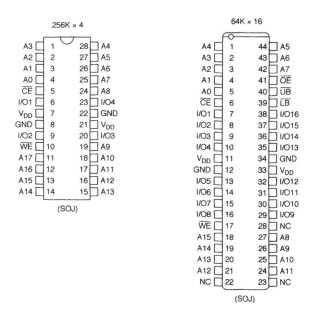

Figure 3.6   Pinouts of two 1M SRAMS with different bus widths (source: Toshiba [2])

3.4   Fast Technologies

3.4.1   *BiCMOS Technology for Speed*

The addition of bipolar drivers internal to a CMOS SRAM gives additional speed due to the higher drive current.

Some timing parameters for a typical 64K 8K×8 SRAM in the same CMOS technology with and without the addition of internal BiCMOS drivers are shown in the table Table 3.1 [2].

As tighter technologies are used for the same density SRAM, the timing parameters for both CMOS and BiCMOS SRAMs improve. Since the cost of the BiCMOS circuitry increases the cost of the SRAM as well as the speed, such parts tend to be used in higher end computer applications, such as workstations, where the speed is required.

If the bipolar circuits used are ECL (Emited-Coupled-Logic), then there can be a timing delay due to the need for level conversion in an SRAM with CMOS I/O levels.

One proposal for dealing with this problem in clocked SRAMs is to pipeline the internal architecture of the SRAM (Clocked SRAMs will be discussed further in a later section.) Pipelining involves sectioning the internal datapath of the SRAM into parts with approximately equal signal delay. The clocked cycle for a pipelined part is equal to the delay of its longest section, so the cycle time for the part is improved. This is illustrated in Figure 3.7(a) for an experimental CMOS SRAM which proposed ECL for the sense amplifiers and then level shifts back to CMOS outputs. The figure shows the clock cycle time of the SRAM both non-pipelined and pipelined.

The cycle time of the pipelined part is faster because the ECL-CMOS converter section is clocked separately. The drawback is in the extra cycle of latency added by the internal pipelining.

There are also very fast bipolar ECL SRAMs where the interface circuitry is also bipolar. These tend to be very fast small parts used in applications such as scratch pad and buffer storage.

An example is a 4.0 ns 4K×4 ECL SRAM with separate inputs and outputs. The simple write timing diagram is shown in Figure 3.7(b).

ECL I/O SRAMs have been offered in two different variations of the interface – 10K and 100K ECL. The different ECL interfaces will be discussed in Chapter 8.

Table 3.1   Comparison of CMOS and BiCMOS SRAM characteristics

	CMOS	BiCMOS
Timing parameters (ns)		
t_{RC} read cycle time	15	10
t_{ACC} address access time	15	10
t_{OE} output enable access time	9	6
Active current. (mA) (at 15 ns cycle)	135	155
Standby current (mA) (inputs at rails)	1	10

Source: Toshiba.

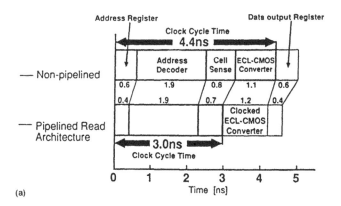

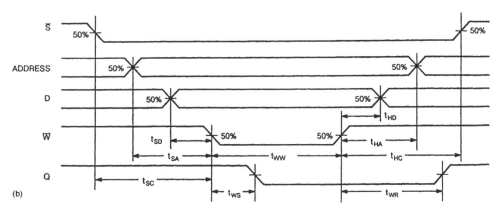

Figure 3.7 Illustration of use of ECL circuitry on fast SRAMS: (a) comparison of clock cycle for non-pipelined and pipelined architecture using internal ECL circuitry (source: K. Yokomizo and K. Naito (1993) [11], permission of IEEE); (b) write timing diagram for a fast separate I/O ECL SRAM (source: Cypress [1])

A drawback of the ECL SRAMs is the higher power dissipation at the speeds run. The power dissipation of the part used in this example running at speed is over 1.5 W. The cost of the technology is also higher than CMOS. For these reasons, only very small ECL SRAMs are used in applications where the speed is essential and justifies the cooling required and the cost.

3.4.2 GaAs Technology for Speed

GaAs SRAMs are very fast with speeds in the picosecond range. They are, however, difficult to manufacture and to scale. As a result, only very small GaAs SRAMs tend to be used in applications where the requirement for speed or for radiation hardness justifies the cost and difficulty in obtaining density in the material.

3.5   Effect of Lower Power Supply Voltage on Speed

If a CMOS SRAM designed for 5 V operation is run at 3.3 V, it will operate, but the operation will be slower than at 5 V. If, however, a CMOS SRAM is designed for 3.3 V operation, and in a technology appropriate for 3.3 V operation, then the speed should not decrease.

An example of the speed decreasing as the power supply voltage is dropped is a 64K×16 SRAM [4] in 0.8 micron CMOS technology which is specified at 15 ns access time at 5 V, but at 25 ns access time at 3.3 V.

The characteristic derating curve for change in access time with power supply voltage is given by some suppliers in the datasheets for the parts. For example, a derating curve for normalized access time vs. power supply voltage for a slow 32K×8 CMOS SRAM is shown in Figure 3.8(a). This figure indicates a 20 percent change in access time is indicated across the specified operating range of the part at 258C.

Power dissipation, however, decreases as the power supply level is dropped, giving the high speed SRAM an advantage when designed for the lower voltage. An example is plotted in Figure 3.9 [2], for a 1M CMOS SRAM.

When the 1M CMOS SRAM is operated at 5 V power supply at 15 ns cycle time, the maximum operating current is 260 mA, whereas for a 1M CMOS SRAM operated at 3.3 V power supply also at 15 ns cycle time, the maximum operating current is 200 mA. The power is therefore significantly less (23 percent less) for the same operating speed at a lower voltage.

3.6   Effect of Temperature on Speed

For a CMOS SRAM, as for CMOS logic in general, the colder the circuit is the faster it will run. At least one supercomputer company even tried dipping their circuitry in liquid nitrogen to exploit this phenomenon. A chart of access time vs. ambient temperature is shown in Figure 3.8(b).

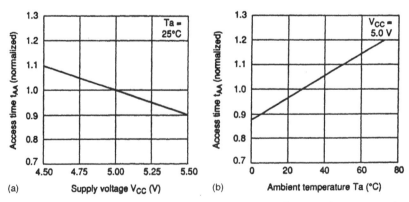

Figure 3.8   Typical characteristic curves for CMOS SRAMS: (a) access time vs. supply voltage; (b) access time vs. ambient temperature (source: Hitachi [4])

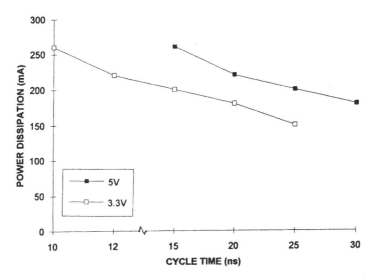

Figure 3.9 Power dissipation vs. cycle time for two operating voltages (3.3 V and 5 V 64K×16 CMOS SRAM) (source: Toshiba [2])

Other experiments have shown that the speed derating curve continues to drop to liquid nitrogen temperatures.

3.7 Revolutionary Pinout for Speed

Notice that both the 256K×4 and the 64K×16 in the example in Section 3.3.3 are 1M SRAMs and are made by the same manufacturer in the same technology. In spite of this, the ×16 is specified as 3 ns slower than the ×4 organized part.

This loss of speed is due to a phenomenon called ground bounce or ringing. When an output of an integrated circuit switches rapidly from one voltage level to another, a surge of current occurs in the self inductance of the ground and power supply connections setting up an inductive ringing effect which is then damped out over time. This is called the $L\frac{di}{dt}$ effect.

It is necessary to wait until the output is at a stable voltage level to be able to read it. Each output that swings adds to the current spike so that a device with many outputs has a larger ringing effect than one with fewer outputs and it takes longer before the data can be read. Multiple power and ground pin can reduce the ringing but add to the size and cost of the package.

The ground bounce effect was particularly a problem with the traditional SRAM package which had a ground at one end and a power supply input at the other. A newer configuration, called the "revolutionary pinout", with center power and ground pins tends to have less ground bounce both because the self-inductance is less in the shorter wires but also because the mutual inductance effect of having the VSS and VDD together opposes the self-inductance effect and reduces the bounce.

For a more detailed discussion on the ground bounce effect and center power and ground solution see Reference 3.

Pinouts of a 128K×8 SRAM with end power and grounds (conventional) and with center power and ground (revolutionary) is shown in Figure 3.10(a).

The chip layout possible with the center power and ground configuration also tends to allow a quieter chip design. The power bus goes to the center of the chip bounded by I/O's and then control pins. The addresses go to arrays which can be on the ends of the chips away from the noise at the chip center.

For reduced noise in the array from output ground bounce, it is also possible to use a different set of power and ground pins for the array and for the outputs.

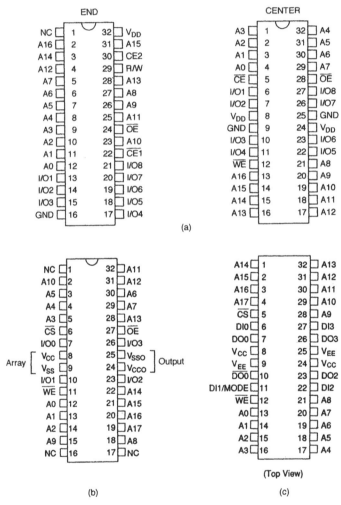

Figure 3.10 Illustrations of revolutionary pinout for fast SRAMS: (a) comparison of end and center power ground for 128k×8 SRAM (source: Hitachi [4]) (b) separate power/ground pins for array and I/O for 256K×4 SRAM; (c) ECL interface for a 256K×4 SRAM (source: Toshiba [2])

Figure 3.10(b) shows a fast 256K×4 SRAM with the revolutionary pinout having a set of VSS and VCC for the array and a set of VSSO and VCCO for the outputs.

3.7.1   *Revolutionary Interface on ECL SRAMs*

For SRAMs in very high speed applications made with ECL compatible interfaces, the revolutionary pinout is also applicable. The input and output voltage swing is limited with an ECL compatible interface so that the data may be read more rapidly.

The revolutionary pinout of a 256K×4 SRAM with 100K ECL interface using separate I/O is shown in Figure 3.10(c) [4].

The block diagram for the circuit showing the output read/write circuit block for the separate I/O is shown in Figure 3.11. The separate I/O adds to the speed.

The major improvement in transition speed is in the transition rise and fall times for the low swing ECL interface. The rise and fall times for a comparable access time CMOS TTL I/O SRAM from the same manufacturer are 4 ns, whereas the rise and fall time for the ECL I/O SRAM is 1.5 ns.

A comparison of the voltage and interface levels for ECL and TTL memories is given in the discussion in Chapter 8.

3.8   Latched and Registered SRAMs

3.8.1   *Overview*

The addition of a latch or a double latch called a register on the inputs or outputs of a SRAM can improve the system control of the memory. Figure 3.12(a) illustrates

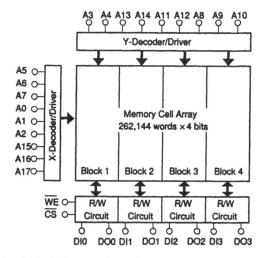

Figure 3.11   Functional block diagram for a fast separate I/O SRAM (source: Hitachi [4])

schematic symbols for a non-inverting buffer, a latch, and a register. Further discussion of the difference between a latch and a register is given in Reference 3.

3.8.2   Latches

Consider an SRAM with no control pin other than the write enable, which is required on any "read-write" memory. The SRAM is totally asynchronous, that is, the address logic levels are applied to the address pins and after a set-up time (t_{AS}) are transferred into the chip. (Also, at a time (t_{OH}) after the address transition the outputs go to an indeterminate state.) After the maximum address access time (t_{AA}) has elapsed for a read, the data becomes valid on the output as shown in Figure 3.12(b).

An example of an output latch is the control pin used as an output enable function which was discussed previously. The processor was able to access the data rapidly from the output latch by not opening the latch until ready for the data.

A typical example of an input latch is the chip select (CS\) control on the data lines of an SRAM. Figure 3.13(a) shows the block diagram of a separate I/O SRAM with a chip select controlled latch on the input and output data lines.

Access time from chip select is faster than access time from address since the address hold time (t_{AS}) is eliminated as shown in the write timing diagram in Figure 3.13(b). This is due to the fact that the array has already been addressed and the chip enable latch just controls the data flow. Write occurs during the overlap of a low CS\ and a low WE\.

If the I/O's are common as shown in the block diagram in Figure 3.14, then it is necessary to include also the time to turn the outputs off before applying data to the inputs.

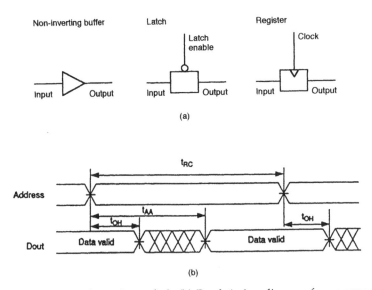

(a)

(b)

Figure 3.12   (a) Various schematic symbols (b) Read timing diagram for an asynchronous SRAM (Source: Motorola [8])

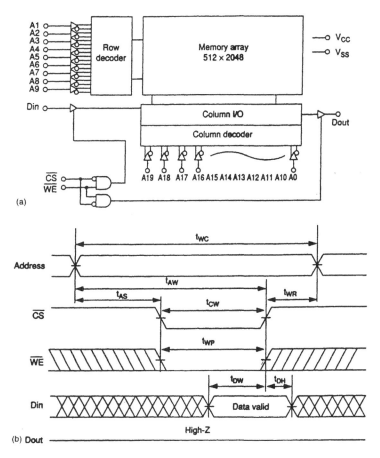

(a)

(b) Dout

Figure 3.13 Separate I/O SRAM showing chip select: (a) block diagram; (b) CS\ controlled write timing diagram (source: Hitachi [4])

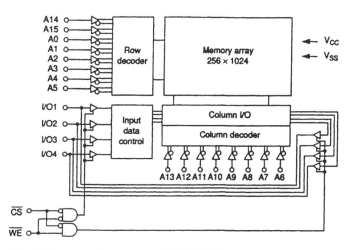

Figure 3.14 Block diagram of a common I/O SRAM with chip select (source: Hitachi [4])

The difference between a separate and common I/O SRAM is illustrated in the write timing diagram in Figure 3.15(a) for a separate I/O SRAM controlled by write enable WE\. Here the data can be valid on the input before the data output has turned off.

In Figure 3.15(b), however, which is a write timing diagram of a common I/O chip, the data out must be off and the I/O line at high impedance before the input data can be applied. The time to turn the output off (t_{WZ}) is a constraining factor for the common I/O part but not for the separate I/O SRAM. The separate I/O SRAM may therefore be faster in a system with frequent back-to-back read and write cycles.

It is also common to have both an output enable latch and a chip select latch on the same SRAM as shown in the block diagram in Figure 3.16(a).

With a CS\ latch, as shown in Figure 3.16(b), the address is transferred to the SRAM on or before the chip select is turned on (CS\ goes low). The data becomes valid on the output after a time t_{ACS} which is the chip select access time. The OE\

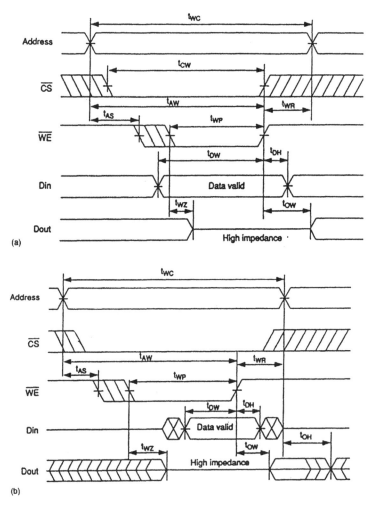

Figure 3.15 Write enable (WE\) controlled timing for (a) separate and (b) common I/O (source: Hitachi [4])

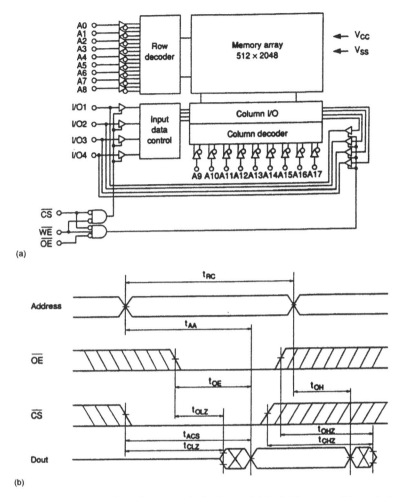

Figure 3.16   SRAM with both CS\ and OE\ latches: (a) block diagram; (b) read timing waveform (source: Hitachi [4])

could be tied low so the output latch remained open and data flowed out after the time t_{ACS}.

If the OE\ latch is used then the output goes to low impedance a time t_{OLZ} after the OE\ latch turns on and after a time t_{OE} the data becomes valid on the output. The CS\ pin could be tied low so the input latch remained open and the addresses flow directly into the SRAM.

3.8.3   Registers

A register is a double latch. In a clocked system the data is taken into the register on one clock, after which time it no longer needs to be held on the input. On the next clock it is passed out of the register.

A register added on the input pins of an SRAM can free the processor from the need to hold the input information on the SRAM for a longer period of time. Input latches can also be added on data, address, or control pins. A register added in the output path of an SRAM can permit the data to be held and clocked in or out of the SRAM when the processor is ready for it.

In the case of a register on the inputs of an SRAM the applied logic level is set up on the first latch of the input register, then latched in on the first clock where it is held on the second latch flowing into the RAM with the data appearing at the outputs after the normal address access time.

If there is also a register in the output path, then the data flows into the input latch of the data register and a second clock latches it into the output latch of the output register where it flows through into the output buffer and after a time t_{AC} appears on the outputs as valid data.

A block diagram of an SRAM with an address register and input and output data registers is shown in Figure 3.17. On one clock of a read cycle the addresses are latched into the address register, the addresses are decoded and the data output appears at the output data register. On the second clock the data is latched into the output register and flows out onto the output buffers. This "registered" SRAM is synchronous to the system clock. It is also our first example of a pipelined synchronous SRAM.

Pipelined in this case means that the various functions are clocked through the SRAM in several clocks. The control pipe, for example, has a strobe register which clocks in the control signals such as OE\, CE\ and WE\ on the first clock. The control signals need to reach the end of the pipeline on the second clock to be in synchronization with the data coming out of the output data register on the second

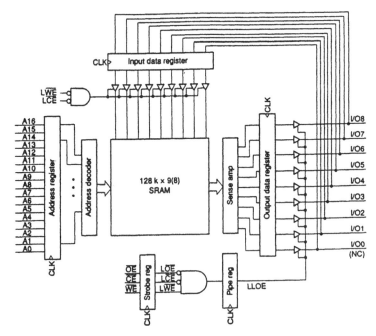

Figure 3.17 Block diagram of a synchronous "registered" SRAM (source: Hitachi [4])

clock. For this reason an extra register is inserted in the control signal pipe whose only purpose is to delay the signal until it is needed. There is also an input data register which latches in the data on the first clock so it is presented to the SRAM along with the address.

We say that this SRAM has a two-cycle read latency, since the data appears on the output after two clocks, and a zero-cycle write latency as shown in the timing diagram in Figure 3.18.

Looking first at Figure 3.18(a) for a pipelined read cycle, the address $A(n)$ and the control signals DE\, OE\, and WE\ are applied to the inputs. After the appropriate set-up times have elapsed, these signals are clocked into the SRAM on clock 1. The control signals are held on the pipe register and the data, after being accessed, is held on the output data register. On the second clock the control signals are applied turning on the outputs and after a time t_{AC}, which is the usual access from CE\, the data $Q(n)$ appears on the outputs.

Meanwhile, the next address $A(n+1)$ and the control signals appropriate for this address are latched into the SRAM on clock signal 2 with the corresonding data then held on the output data registers awaiting clock 3. On clock 3, the control signals are applied, the data is latched onto the second latch of the data register and appears on the outputs after a time t_{AC}.

This forms a filled pipeline within the SRAM which permits a series of data to be accessed in a time which is shorter than the normal address access time after the first access. The penalty that is paid is the wait for the first access which is called the latency and is measured normally in clock cycles. In the case of this SRAM, the read latency is two clock cycles.

The write latency, however, is zero as shown in Figure 3.18(b). This is because, after the addresses and control signals have been latched into the SRAM, the write proceeds independently of the system. From the point of view of the system, the latency is zero. The write access time from clock is therefore t_{IH}, the required hold time of the addresses and control signals after clock.

Registers have the advantage of adding flexibility in the system since the processor is not required to hold the data on the SRAM for longer than it takes to latch it in, nor is it required to wait for the output longer than it takes to retrieve the data from a latch. The pipeline of fast data can also add speed to a system that does not need to alternate reads and writes frequently.

Latched and registered SRAMs are also referred to as self-timed SRAMs, ST-SRAMs, clocked SRAMs or synchronous SRAMs since they are under the control of the system clock.

For example, a 256K×4 SRAM without latched inputs and outputs might have an access time of 12 ns. With the addition of latches on the inputs and outputs, the access time from clock is 6 ns [4].

Many SRAMS which are called "synchronous" are not fully synchronous but have some pin that is still asynchronous. For example, the synchronous SRAMs used with the 586 class processors have an asynchronous output enable to give increased control over the output. This can be seen in the block diagram and timing diagram in Figure 3.19 [5].

The block diagram in Figure 3.19(a) shows the input registers on the address lines, the data in and out lines (DQ), the chip enable (E\), and the write enable (UW\,

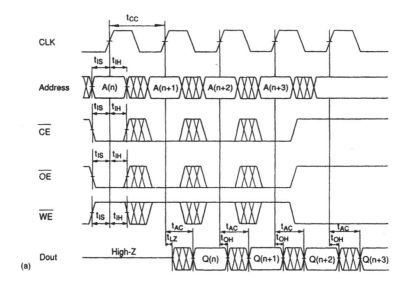

(a)

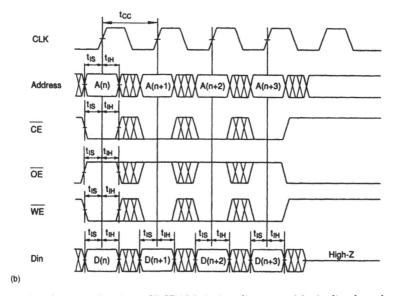

(b)

Figure 3.18 Synchronous "registered" SRAM timing diagrams: (a) pipelined read cycle; (b) pipelined write cycle (source: Hitachi [4])

LW\) lines. There is no input register, however, between the output enable (G\) and the outputs. The G\ latch simply turns the outputs on and off as required by the system. This can be seen in the timing diagram in Figure 3.19b which shows a read followed by a write. The registered inputs (A, DQ, W\, ADSP\) are clocked in by the

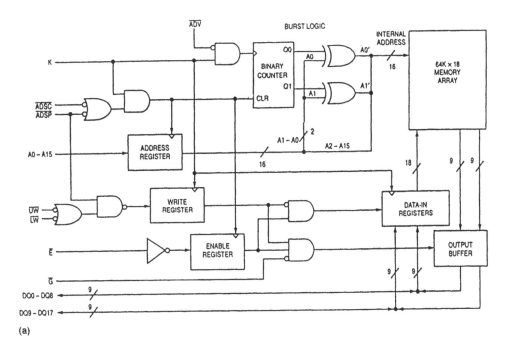

(a)

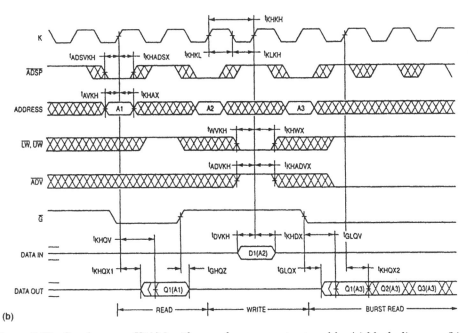

(b)

Figure 3.19 Synchronous SRAM with asynchronous output enable: (a) block diagram; (b) read/write timing diagram (source: Motorola [5])

system clock. The output enable (G\) is used to turn the output off after the read, and turn it back on after the write is complete.

Observe that the notation for the pins and the timing parameters are different in this diagram than in the others seen so far in this book. The notation used in this figure for the symbols is the JEDEC standard, and that used for the timing diagrams is the IEC/IEEE Standard timing notation. Unfortunately a system of pin identifiers was established by common practice before the JEDEC symbols were established, and they are still commonly used. The system of timing notation used up to this point was also established by common usage and is that commonly seen in databooks. Both systems are mnemonic.

3.8.4 *Synchronous (Registered) SRAM with Separate I/O Option*

Another requirement for fast synchronous SRAMs is in pipelined systems, systems with multiple processors or where a local processor has a bus isolated from a common system bus.

All of these applications require the high speed and processor synchronizing capability of the fast synchronous SRAMs. Because of multiple processors or devices attempting to access the RAM, there is also a serious potential for bus contention. This can be avoided either by having multiple identical ports (dual ports) on the SRAM or by having separate inputs and outputs.

The easiest solution is separate inputs and outputs. The block diagram of a fast 128K×9 separate I/O synchronous SRAM is shown in Figure 3.20. All addresses, inputs, outputs, and control signals are registered. In addition, because errors in an SRAM in a multiprocessor system can be serious, the part shown has boundary scan registers which form a serial chain through the inputs and outputs of the SRAM.

For systems with fast streaming requirements or multiple processors requiring full access to the part, a more complex configuration on the SRAM can be easier to use in the system. For this purpose, a 1M synchronous SRAM designed with the option for dual ports or separate inputs and outputs has been developed [5]. The pinout is shown in Figure 3.21(a) and the block diagram in Figure 3.21(b).

The processors and systems which could be interfacing with this SRAM are called in the notation, the "processor" and the "system". Pins PIE\ and SIE\ are the processor and system input enables. Likewise POE\ and SOE\ are the output enables. DQ pins are similarly noted.

The PDQP pin is a data parity pin. Parity involves adding a check bit for every eight I/O bits. It will be discussed in the next chapter. It is frequently used in multiprocessor or workstation environments, although much less common in the PC environment.

The address inputs, write enable and input enables are all synchronous (registered). The data I/O's have input registers triggered by the rising edge of the clock. They also have three-state output latches which are transparent when the clock is high and latched when it is low. This permits two modes of operation.

The part can be used as a dual port SRAM with the PDQs being one port and the SDQs being the other. It can also be used in separate I/O applications where the PDQs and SDQs are treated as either inputs or outputs.

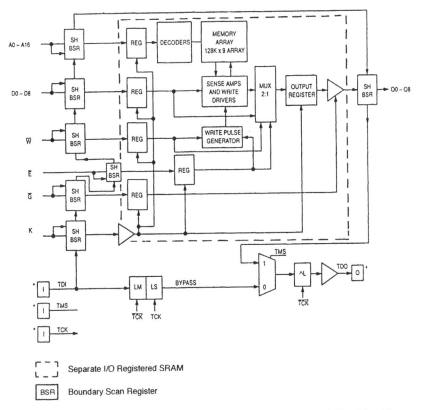

Figure 3.20   Block diagram of a fast 128K×9 separate I/O registered SRAM with external boundary scan chain

This SRAM can also be used in pipeline systems using a "streaming" feature where data is just passed through the RAM unchanged. Streaming is accomplished by latching in data from one port and out through the other port by using the asynchronous output enable as shown in the timing diagram in Figure 3.22(a). A truth table for this rather complex part is shown in Figure 3.22(b).

The majority of fast synchronous SRAMs are used almost entirely in SRAM caches in fast systems. Many of them are specifically tailored for second level cache with specific processors. For this reason, more detailed discussion of the synchronous SRAMs used in caches will be delayed until the next chapter.

3.8.5   *Fast SRAMs with Early Write Feature, Zero Bus Turnaround*

Fast synchronous SRAMs for cache applications are used primarily as "read mostly" memories. When a read burst followed by a write is required, several cycles are necessary to clear the read burst and begin the write.

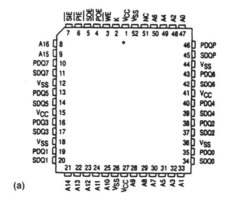

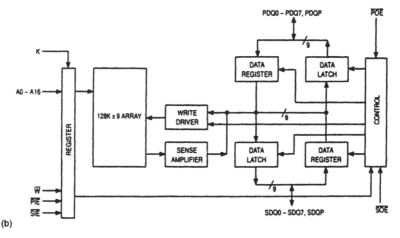

Figure 3.21   128K×9 synchronous dual I/O or separate I/O SRAM: (a) pinout; (b) block diagram (source: Motorola [5])

For applications such as switching systems, which also require fast SRAMs, the bus must be able to turn immediately from a read to a write cycle and from a write to a read. Several SRAM manufacturers developed SRAMs with the ability to do fast read-write and write-read. Called variously "late write", and "zero bus turnaround", the features of these SSRAMs increase the effective bandwidth of the SRAM in read/write situations.

"Zero Bus Turnaround", developed by IDT, saves a clock cycle when the bus is turned from a read cycle to a write cycle or vice versa, greatly increasing system speed in the worst case. Using this architecture, synchronous SRAMs can mix read and write operations in any combination with only a single clock cycle required per operation.

The "Late Write" SSRAM was developed by Motorola. It involves delaying the write data by one cycle. Write data is normally coincident with the address in a write cycle for an SSRAM.

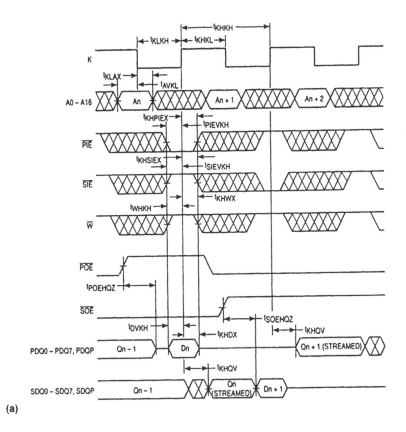

(a)

(b)

$\overline{W}$	$\overline{PIE}$	$\overline{SIE}$	$\overline{POE}$	$\overline{SOE}$	Mode	Memory Subsystem Cycle	PDQ0 – PDQ7, PDQP Output	SDQ0 – SDQ7, SDQP Output
1	1	1	0	1	Read	Processor Read	Data Out	High-Z
1	1	1	1	0	Read	Copy Back	High-Z	Data Out
1	1	1	0	0	Read	Dual Bus Read	Data Out	Data Out
1	X	X	1	1	Read	NOP	High-Z	High-Z
X	0	0	X	X	N/A	NOP	High-Z	High-Z
0	0	1	1	1	Write	Processor Write Hit	Data In	High-Z
0	1	0	1	1	Write	Allocate	High-Z	Data In
0	0	1	1	0	Write	Write Through	Data In	Stream Data
0	1	0	0	1	Write	Allocate With Stream	Stream Data	Data In
1	0	1	1	0	N/A	Cache Inhibit Write	Data In	Stream Data
1	1	0	0	1	N/A	Cache Inhibit Read	Stream Data	Data In
0	1	1	X	X	N/A	NOP	High-Z	High-Z
X	0	1	0	0	N/A	Invalid	Data In	Stream
X	0	1	0	1	N/A	Invalid	Data In	High-Z
X	1	0	0	0	N/A	Invalid	Stream	Data In
X	1	0	1	0	N/A	Invalid	High-Z	Data In

Figure 3.22   128K×9 synchronous dual port SRAM: (a) streaming timing cycle; (b) truth table

3.8.6   *Double Data Rate Synchronous SRAMs*

The next step in gaining speed with SSRAMs was the Double Data Rate (DDR) synchronous SRAM. A DDR synchronous RAM runs the data bus at twice the speed of the clock. This is accomplished by clocking data on both the uptick and downtick of the system clock, which effectively doubles the speed of the data.

The doubled rate of data reaching the output buffer of the SSRAM is accomplished by using a wide internal bus which obtains two words of a burst of data on every clock cycle. A register which can store the two words of data is on the output data bus of the SRAM, followed by a multiplexing circuit which interleaves the two bits of data out at double the clock rate.

For example, a 2M-bit DDR SSRAM today is capable of 225 MHz to 250 MHz data rate. Because of the high data rate, such an SSRAM would probably use the low swing HSTL interface and be packaged in a low capacitance Ball Grid Array (BGA) package. A typical DDR SSRAM uses a differential clock and the data bus is clocked at twice the frequency of the main clock. The write occurs a cycle after the write address is given. This gives the device a write latency of "1" making this a "late write" SSRAM device.

3.9   FIFOs

FIFOs are fast dual port SRAMs made in various configurations for use as buffers for multiprocesser and peripheral communications, for serial communications networks, and for fax and modems. Figure 3.23 shows block diagrams of various types of SRAM FIFO buffers [7].

Figure 3.23(a) shows a ×9 asynchronous FIFO with retransmit capability which allows the read pointer to be reset to its initial position. The part is used in multi-processing and rate buffering types of applications.

Figure 3.23(b) shows a 64K×4/5 parallel FIFO. The shift-out (SO) signal causes data at the next to last word to be shifted to the output while all other data shifts down one location in the stack. Reading and writing are asynchronous so the FIFO can be used as a buffer between two digital systems with widely varying operating frequencies.

Figure 3.23(c) is a ×9 parallel–serial FIFO with 50 MHz serial input and output frequency. It performs all combinations of serial and parallel operations and has asynchronous read and write operations. It is used in a range of applications including high speed data acquisition systems, LAN buffers, modem data buffers, FAX raster video data buffers, laser printer data buffers, magnetic media controllers and high speed parallel bus-to-bus communications.

Figure 3.23(d) shows a ×9 synchronous FIFO with clocked read and write controls. The input port is controlled by a free-running clock and two write enable pins. Data is written in on the rising clock edge with write enable asserted. Another clock pin controls the output port along with two read enable pins. This FIFO is used for data buffering in graphics, LANs, and multiprocessor systems.

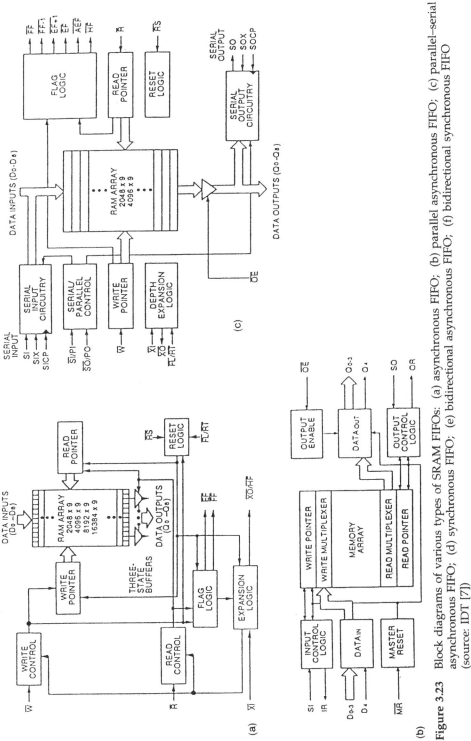

Figure 3.23 Block diagrams of various types of SRAM FIFOs: (a) asynchronous FIFO; (b) parallel asynchronous FIFO; (c) parallel-serial asynchronous FIFO; (d) synchronous FIFO; (e) bidirectional asynchronous FIFO; (f) bidirectional synchronous FIFO (source: IDT [7])

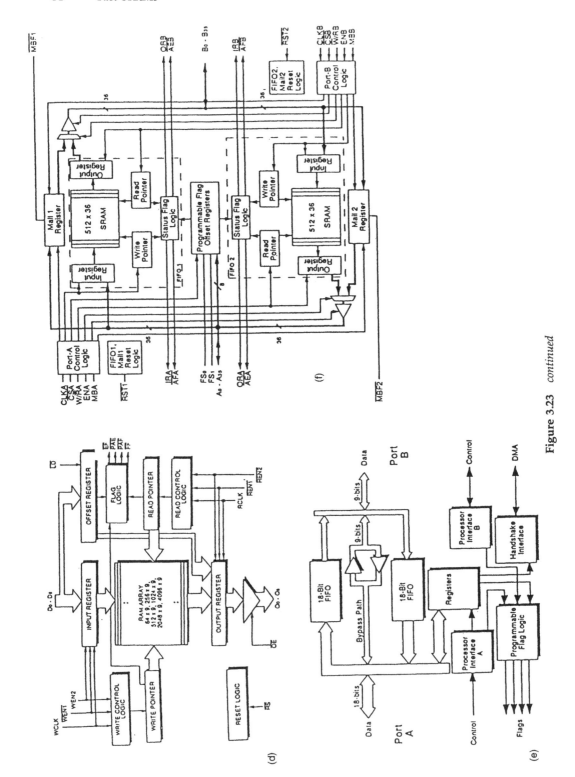

Figure 3.23 *continued*

Figure 3.23(e) shows a bus-matching FIFO for asynchronous bidirectional transfers between processor and processor and between processor and peripheral.

Figure 3.23(f) shows a synchronous bidirectional FIFO that runs at 67 MHz with read access times of 11 ns. The two independent FIFOs buffer data in opposite directions; in addition communications between each port may bypass the FIFOs via the two mailbox registers.

3.10 Fast Non-volatile Memories

Floating gate non-volatile memories include EPROM, EEPROM, and Flash memories. These parts can have a fast read operation by using high speed techniques such as a synchronous interface.

Because of the delay that is inherent in charging and discharging the floating gate that is used for data storage, the write and erase operations tend to be slow. A technique used for making the erase and program operations of a floating gate non-volatile memory appear fast is to integrate a small SRAM register on the input and output structures. Several words of data can be prefetched on a wide internal bus, stored in the register, and fed out at a higher rate of speed than that inherent to the memory.

Bibliography

1. *High Performance Static RAM Databook*, Cypress, 1994.
2. *Static RAM Data Book*, Toshiba, 1994.
3. Prince, B., *Semiconductor Memories*, 2nd Edition, John Wiley and Sons, 1992.
4. *SRAM Databook*, Hitachi, 1993.
5. *Fast Static RAM Databook*, Motorola fourth quarter 1993.
6. Avoiding bus contention in fast access RAM designs, Application Note AN971, Motorola Semiconductor, *Fast Static RAM Databook*, Q395.
7. Specialized memories and modules, *IDT Databook*, 1994.
8. Peters, M., Motorola, private correspondence.
9. *Fast Static RAM Component and Module Data*, Motorola, Q395
10. *High Performance Static RAM Databook*, I.D.T., 1994
11. Yokomizo, K. and K. Naito, Design Techniques for High-throughput BICMOS Self-timed SRAMs, Journal of Solid State Circuits, Vol. 28, No. 4, April 1993, pp 484.
12. *Choosing the right SRAM*, Technical Note TN-05-25, Micron Technology, February 1998.

4 Fast Cache Memory

4.1 Overview

Computer systems have suffered for the last 20 years from a mismatch in speed capability between the processor and the memory. This first became serious for higher end systems such as mainframes and workstations. As a result, these systems were the first to use SRAM cache. SRAMs bridged the speed gap by providing a small amount of high speed, but expensive, memory cache between the processor and the DRAM. The SRAMs have been positioned as close as possible to the processor to avoid transmission line related speed loss. Various architectures of SRAMs have been used to improve the efficiency of the cache, and fast technologies such as bipolar and BiCMOS have been used to increase speed.

In the past few years the trend has been for cache subsystems to move downstream into high end commodity PCs. The high growth rate of the PC market, has resulted in a requirement for standard, lower cost cache. Figure 4.1 shows a projection for the growth rate of the PC cache subsystem market split by size of cache. Subsystems are shown with both external L2 cache and internal L1 cache.

The capability of integrating a first level of cache onto the processor has been followed by new generations of faster, higher density second level cache made in standard configurations and by many suppliers. The addition of a synchronous, or self timed, interface and pipelined architecture has allowed high density CMOS SRAMs to run at the speed required for second level cache in the new PC systems.

These cache SRAM product trends are discussed in this chapter.

4.2 Cache Concept and Theory

The cache concept is simple, but cache theory is much more complex and several excellent books already exist on this subject [1,2]. The cache concept will be discussed

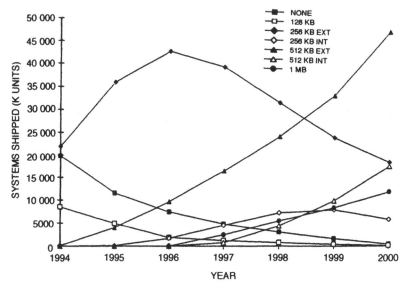

Figure 4.1   Shipments of systems with different size caches (thousand units) (source: Memory Strategies International [19])

here along with very simple cache theory as it affects the properties of the SRAMs used in caches.

4.2.1   The Problem

The problem that led to the use of caches is that processors in computers would like to get data faster than reasonably priced memory banks are able to supply it.

Modern commodity processors are able to run 100–200 MHz internally. Common bus widths are 64 bits (8 Bytes). That means that a single scalar processor wants data to arrive at the rate of 800MB/sec to 1.6GB/sec. If the processor is superscalar, that is, it processes more than one instruction in a cycle, then it can accept data in multiples of these datarates up to 3GB/sec.

For example, if we look at the raw internal speed currently claimed for the various processors, the DEC Alpha chip claims 175 MHz and the Pentium claims 133 MHz. The memory bus, however, tends to be run at a fraction of the internal speed of the processor. The 175 MHz Alpha and 133 MHz Pentium run their memory buses typically at 50–60 MHz. Even 50 MHz, however, is fast for a DRAM or a commodity SRAM as shown in the chart in Figure 4.2.

Commodity asynchronous CMOS SRAM access times are still less than 50 MHz, possibly reaching 80 MHz by the end of the decade. Commodity DRAM random access times are far slower. With fast hyperpage (EDO) mode the fast DRAMs are expected to be capable of 80 MHz and with synchronous interfaces up to 200 MHz. Synchronous SRAMs, on the other hand, are expected to be capable of speeds up to 250 MHz and DDR SSRAMs of speeds of 500 MHz or higher.

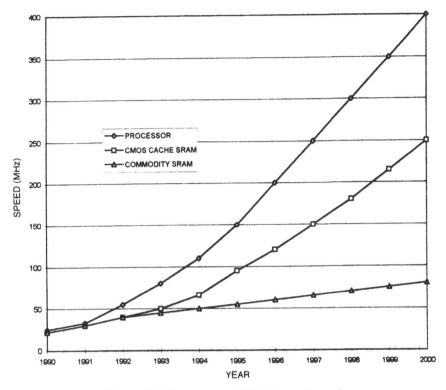

Figure 4.2   Processor and SRAM speed trends

SRAMs clearly will continue to be faster than DRAMs for the foreseeable future.

The second problem that system designers face is that SRAMs cost more than DRAMs. The system designer must therefore make a trade-off between the optimal speed capability of the system and the cost of the system. The lowest cost route is to use DRAMs if possible.

4.2.2   Supplying Data from DRAM Main Memory

Several considerations arise in attempting to supply the processor with its required datarate from DRAM main memory. High density commodity DRAMs are normally used for main memory in computing systems. A typical speed for DRAMs operating in EDO mode is about 50 MHz and for SDRAMs about 100 MHz.

Assume the required main memory in a system is 16MB. On a 64-bit bus, then, the system memory would be configured as 2M×64. This can be supplied by eight 2M×8 DRAMs which can supply data at the rate of 8MB × 50 MHz (400MB/sec). This is half of the 800MB/sec minimum that a simple scalar processor would like to see.

There are several possible solutions for speeding up the average rate at which data arrives at the processor from main memory. One possibility would be to have two banks of four 1M×16 DRAMs which alternate in supplying data to the computer, a process called interleaving. With the interleaved banks, the data is supplied almost twice as fast as before, so the datarate is nearly 800MB/sec. It will not be quite that fast since the logic required in the system to multiplex the banks of DRAMs delays the signal and the timing will be more complex. There will also begin to be transmission line effects even at 50 MHz which will slow the system slightly.

The cost for the two-bank interleaved main memory will also be higher. The wider DRAM will use more silicon and require a package with more leads and possibly a larger package. Cost in DRAMs is frequently calculated in terms of silicon space and cost per pin of the package. The additional board used in the interleave will also add cost and occupy additional space. Fast logic will need to be added for the interleave, adding to the cost.

Another possibility is using the new fast DRAMs with burst accesses. These still do not reach the effective random access speed of an SRAM due to the longer initial latency of the DRAM.

4.2.3   Supplying Data from a Cache Hierarchy

There is another possibility to get data at the required speed without having to pay the price of main memory SRAMs, which is at least four times the cost of a comparable density DRAM. This is to use fast SRAM, but not use very much of it.

It is possible to divide the memory into a hierarchy of levels called a cache hierarchy which uses a small amount of expensive fast SRAM and a larger amount of low priced commodity DRAM and still achieves the required performance.

The cache is a small fast memory located close to the CPU that holds the most recently accessed data [3]. Frequently the cache is broken into more than one level and the first level is actually integrated into the processor as shown in the schematic block diagrams in Figure 4.3. The second level cache is off the processor chip and can be either on the same bus as the main memory, as shown in Figure 4.3(a), or it can be on a separate bus as shown in Figure 4.3(b).

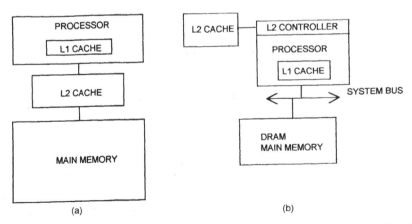

Figure 4.3   Diagram of a traditional cache hierarchy; (b) L2 cache with separate bus (source: M. Peters [17])

The theory of the cache hierarchy is that the data that will be next requested by the processor can be contained in a small and very fast memory that sits next to the processor. This memory needs to be small to enable it to be close enough to the processor to avoid transmission line effects, and small enough to reduce the cost. In addition a small memory tends to be faster than a larger memory.

If this small fast cache can supply the data at a sufficient rate that the processor never, or seldom, needs to go to the main memory DRAM bank itself, then the main memory can be made up of slower, less expensive DRAMs and the average speed of the system can still remain high. The cost of the cache hierarchy is therefore increased by the cost of the added SRAM memory and the control logic needed for that memory, and is reduced by the lower cost of the slower main memory configuration.

In fact, many systems use both first and second level cache and fast main memory to increase the effective speed with which data is supplied to the processor. They also slow down the processor's external speed requirements by double clocking techniques. The result is that the speed requirements of cache SRAMs are determined both by the capabilities of the memories and by the requirements of the processors.

4.3 Effective Speed of the Cache Hierarchy

The effective access time of a memory system is the access time of the cache times the probability that the data required by the processor is in the cache, plus the access time of the main memory multiplied by the probability that the processor needs to go to main memory for its data.

When the CPU finds the item that is addressed in the cache, it is called a "hit". When the item is not found in the cache, it is called a "miss". The probability that the required data is in the cache is called the "hit rate" and the probability that the required data is not in the cache is called the "miss rate" (1 − hit rate). In a "miss" the processor must go to the main memory or the next level of cache for the data.

The formula for effective access time is:

$$\text{TAC}_{\text{eff}} = P_{(\text{hit})} \times \text{TAC}_{(\text{cache})} + P_{(\text{miss})} \times \text{TAC}_{(\text{main memory})}$$

The probability of a hit is complex and is well dealt with in several books devoted to this topic [1,2]. It roughly goes up as cache size goes up, as cache complexity goes up, and as the size of the block of data, which can be moved at one time into the cache, goes up.

4.4 First Level Cache

The highest level of the hierarchy is the first level cache which supplies data to the processor at a rate which requires no delays in processor operation; returning to our example, this would be, 1.6GB/sec.

To accomplish this the first level cache must be very close to the processor so that there are no transmission line delays, the bus between the cache and the processor must be wide enough to supply the data at the rate required, and there must be no delays in the interfaces between the two chips. In addition, it must also contain the data that the processor wants.

Some of these requirements are in conflict. The speed of the cache memory and the width of the interface can be traded off. If the cache memory is 8 bits wide, then it must run at 1.6 GHz to give a datarate of 1.6GB/sec. This speed is not possible today with cost-effective memories.

If the bus of the memory is 64 bits wide, to supply 1.6GB/sec, it needs to run at 200 MHz, and if it is 128 bits wide, it only needs to run at 100 MHz. An SRAM, however, with 128 outputs running at 100 MHz will have significant delays due to transmission effects crossing the interfaces and wiring between the two chips unless careful measures are taken to have proper termination. Also, a memory with 128 I/O's will be a large chip in a large package.

Since these I/O's are between the SRAM and the processor, they can be eliminated by integrating the first level of the memory heirarchy onto the processor chip itself. This integration permits a very wide interface to be used between the processor and the cache without noise problems. It also eliminates the large package and by eliminating the I/O buffers significantly reduces the amount of silicon used. The cache SRAM embedded in the processor logic can run at 100 MHz and have a 128 bit wide bus.

The most critical aspect, however, is ensuring that the data that is required next by the processor is in the cache. Otherwise the speed of the cache is irrelevant. Most of this chapter will be spent discussing techniques by which this is accomplished and the SRAMs which are used.

4.5   Limitations in Size of an L1 Cache

A problem arises in the limitations on size of SRAM possible to integrate with the processor on one chip. The hit rate of a simple cache tends to go up as the size of the SRAM goes up to some point which depends on various factors such as architecture, line size, etc. so the cache should reach its optimal size and still maintain sufficient speed. Figure 4.4 shows one result of the average miss rate of a direct mapped cache vs. log of cache size [13].

At the current time, for technology reasons, the most SRAM that can be integrated into a cache on a processor chip is less than 1Mbit (128KB). This is due to the fact that the SRAM needs to be in the same technology as the logic elements of the processor itself or the cost of the processor will be increased significantly. Unfortunately, the SRAM cell in a logic technology is much larger than the cell in a stand-alone memory technology.

Caches ranging from 64Kbits to 384Kbits have been shown in 0.5m high speed microprocessors. This is less than 50KB, too small in many cases to bring the hit rate to an acceptable level.

There are two possible solutions to this problem: (1) the theoretical hit rate of the L1 cache itself can be increased by increasing its complexity, or (2) a second level of

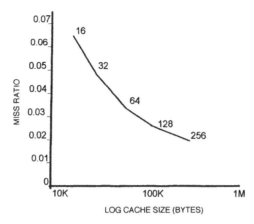

Figure 4.4   Direct map cache hit rate vs. cache size (source: N.P. Jouppi [3])

cache can be added to the hierarchy. To understand solution (1) we need to discuss the theory of the cache in a bit more depth.

4.6   Increasing the Hit Rate of the Cache: Cache Theory

This problem can be solved by taking measures to increase the probability that the data the processor wants next is in the first level cache and the rest of the data is in slower external memory.

4.6.1   *Simple Cache Theory*

A cache's hit rate can be increased by tailoring it specifically to the system. For example, for an L1 cache, the data and instruction cache can be separate. If they each maintain the same bus width the bandwidth of the cache is doubled. In addition, the two can have different configurations and bus widths to provide exactly the speed and bus width required by that part of the processing system. Either one can be direct mapped or partially or fully associative. They can be single or multiple ported. They can be write-through or write-back.

   A broad outline of the cache memory architectures which can be used to increase the hit rate is presented in the following section.

4.7   Cache Architecture

A simple cache consists of a bank of addresses, split into an index and a *tag*, and a bank of data stored in a register close to the processor, as shown in Figure 4.5.

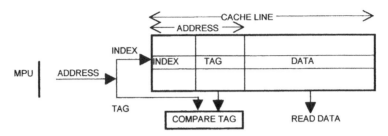

Figure 4.5   Block diagram of a simple cache

The data from a location in the main memory along with its associated main memory address is stored on one line in the cache. The cache locations need to be identified. This is done by taking part of the main memory address (usually the lower order bits) and using these bits to identify a location in the cache. These bits are called the "set" or "index" which is just another name for the cache address.

For example, a location in a 4K-line deep main memory is identified by 12 address bits. If the cache is only 32 lines deep, then a location in the cache needs only five address bits to identify it. The main memory address is therefore broken into a tag consisting of the six higher order bits and an "index" or *set* consisting of the five lower order bits.

The data from the main memory together with the tag part of the address is then stored on a line of the cache corresponding to the index of that address.

The faster the cache can match with the data stored in it, the faster the processor can have its data.

A "line" of the cache consists of an address and the data corresponding to that address. A line is also the minimum unit of information that can be moved between main memory and cache. This minimum unit of information is also called a block.

Recall address can be further split into an index and a tag, with the lower order bits in an address being the index (or set) and the higher order bits the tag. The processor uses the lower order "index" bits to address the higher order tag bits stored in the RAM. (The index bits are the address bits of the cache.) These tag bits are compared with the higher order address bits from the processor. If they match then the data which corresponds to that address is read out of the data RAM part of the cache [3].

In a cache read operation, a controller takes the address which is requested by the processor and breaks it into a set and a tag. It then compares the tag from that address with the tag stored at the set (index) location in the cache. If the tags match a hit occurs, then the corresponding data from that line in the cache is sent on to the processor. If the tags do not match, then the required data is not in the cache and the controller must go to main memory for the proper data.

4.7.1   Principles of Locality of Time and Space

The probability must be high that the requested data will be in the cache or the situation would be worse than just going to main memory to begin with.

What makes this probability high are two principles called locality of time and space. Temporal locality says that information that the processor wants now has a high probability of being needed again soon. Spatial locality says that information that the processor wants next has a high probability of being near the information it wants now.

A cache hierarchy makes use of these principles by transferring to the cache from main memory both the data currently requested and other data lying physically near that data. The larger the block that is transferred, the better the cache takes advantage of locality of space. The more blocks (lines) that can be saved in the L1 cache at one time, the better the cache takes advantage of locality of time.

4.8   Data and Instruction Caches

The information that the processor wants normally falls into instructions and data. The two types of information can be stored in one cache or two.

A system can have a split cache with one for instructions and one for data. The two caches can have different structures to optimize their function. A cache which contains both instructions and data is called a "unified cache".

The block diagram in Figure 4.6(a) illustrates a system with a split cache, and Figure 4.6(b) shows a unified cache.

The data and instruction cache for a first level cache on the processor chip are frequently constructed to optimize the performance of each. They may have different levels of associativity and different widths.

Similarly separate data and instruction caches can be used for external L2 cache. These can be on two different chips or on the same chip.

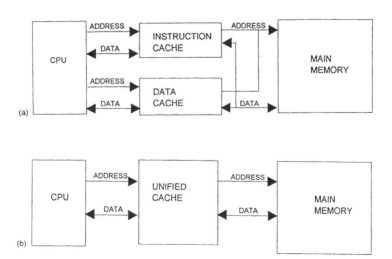

Figure 4.6   Block diagrams of two types of cache architectures: (a) separate data and instruction cache; (b) unified data and instruction cache

An example of a split cache in a single SRAM chip is shown in Figure 4.7 [5]. This part is organized as two independent banks of 8K×20 bits each with associated address latches and control signals. In this case both the data and the instruction cache are identical.

The processor access to the two banks is handled by two synchronous control pins ALE1 and ALE0. The address signal from the processor for array "0" is latched in on the rising edge of ALE0 while ALE1 is inactive low. The CPU can then execute a read cycle on bank 0. Bank 1 is then accessed by latching an address at the rising edge of ALE1 while ALE0 is inactive low.

4.9   Cache Associativity

Different cache memory architectures have different levels of a property called asso-ciativity. The hit rate of the cache can be improved by increasing its associativity. At

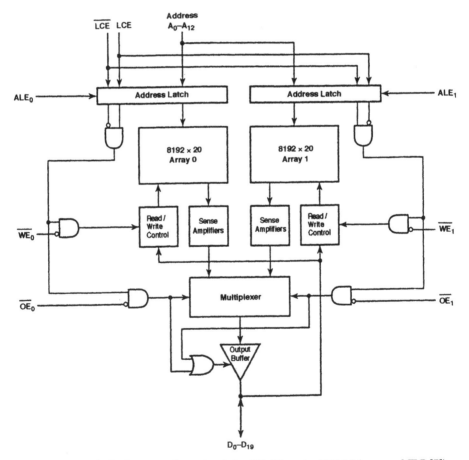

Figure 4.7   Block diagram of two-bank by 8K×20 cache SRAM (source: NEC [5])

the lowest level of associativiy is the direct mapped cache,followed by increasing levels of set associativity. A fully associative cache is called a CAM or Content Addressable Memory. Associativity reflects the number of lines of data being checked for a hit on one cache access. One line of data can be checked with one comparator (direct mapped). Two lines of data can be checked with two comparators simultaneously (two way set associative), etc.

4.9.1 Direct Mapped Cache

A simple, one-bank cache and tag architecture with a single comparator for matching the address requested by the processor is called a direct mapped cache. A block diagram of a direct mapped cache is shown in Figure 4.8(a).

In this case the data from each location in main memory is copied into only one line in the cache along with its tag address bits. The line of the cache where it is stored is identified by its index or set bits. Recall that the full address bits of the location in main memory are broken into higher order bits which are called the tag, and lower order bits corresponding to the depth of the cache which are the index or set (address in the cache) [1].

The direct mapped cache has one comparator which compares the higher order bits of the address generated by the processor with the TAGs in all the locations of the cache.

4.9.2 N-Way Set Associative Cache

If the hit rate of the cache can be improved then the cache is more efficient. An N-way set associative cache improves hit rate of the cache by breaking the cache into "N" identical banks with "N" comparators. Each of the "N" banks has the same set locations. The tag comparison for all the banks is made simultaneously and if the tag (higher order address bits) matches for one of the banks, then there is a hit. It is possible that the same location in main memory can be copied into more than one bank of the cache.

A diagram of a two-way set associative cache is shown in Figure 4.8(b).

The tags from the first bank and from the second bank located at the address set (index) must be compared with the address tag before a match decision is made.

Similarly if the cache is split into eight banks with eight comparators, then the cache is called eight-way set associative.

4.9.3 Content Addressable Memory

A fully associative cache has a separate bank and comparator for each address (tag) and data location. In this case all locations of the cache are compared simultaneously and it is possible that the same address from main memory could be stored in every location in the cache. Figure 4.8(c) illustrates a content addressable memory showing a comparator for each address.

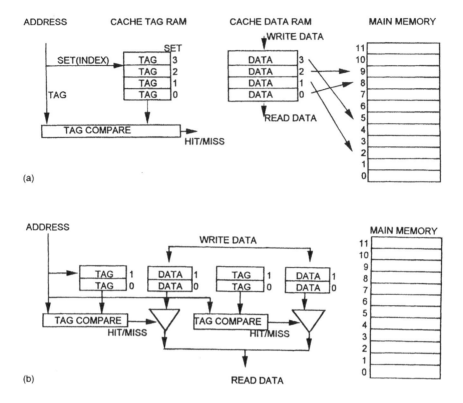

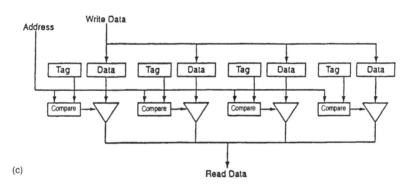

Figure 4.8　Block diagrams of different cache architectures: (a) direct mapped cache; (b) two-way set associative cache; (c) content addressable memory showing comparator for each address (source: S.A. Przybylski [2])

A content addressable memory can also be made with a special nine-transistor cell [11]. This results in a very large expensive cell which is normally only manufactured with a very few data storage bits for cost reasons.

First-level cache that is integrated on the processor chip is frequently made as a content addressable memory since the low density of cells required permits a very large cell to be used.

4.10   Dual-port Caches

SRAMs are also designed with multiple ports for use in multiprocessor systems with shared cache. This permits both of the processors to have direct access to the SRAM.

An example is a 16K dual port RAM from IDT. This part provides two ports with separate control, address and I/O pins that permit independent access for read or write to any location in the SRAM [7]. These parts are available both with interrupt capability and without. A functional block diagram of the part with interrupts is shown in Figure 4.9. It is called "dual-ported" rather than "multi-ported" since the two ports are essentially identical.

There is a shared memory array along with arbitration and interrupt logic. Addresses, I/O's and interrupt control pins for the independent left and right ports support fully asynchronous operation from either port. There are also busy indicator pins and output enable pins for each port. The interrupt flags provide for communications between the ports by allocating a space in memory for a message entered between the ports.

The separate chip enable (CE\) for each port controls power-down circuits that let that port go into standby mode when it is not selected. When a port is selected, it controls access to the entire memory array.

The BUSY\ flags are provided for the case when both ports simultaneously attempt to access the same memory location. The arbitration logic will determine which port has access and set the active BUSY\ flag for the delayed port. An address match or chip enable match can be resolved down to a minimum time of a few nanoseconds. The delayed port will have access when the BUSY\ signal goes inactive.

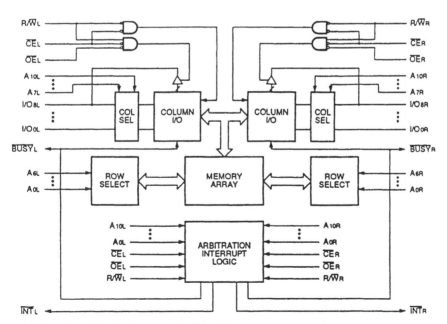

Figure 4.9 Functional block diagram of a 2K×8 dual port SRAM with interrupts (source: IDT [7])

When more than one dual port RAM is used in a memory bank, there is a potential problem related to the existence of arbitration logic in each RAM.

In the case of a contention situation one RAM could lock out the left port and another could lock out the right port with the result that neither side of the bank could be accessed. This can be avoided by linking the bank in such a way that the arbitration logic in one of the RAMs has master control and the other RAMs are "slaves". This can be accomplished by delaying the write pulse to the "slave" RAMs by the maximum arbitration time of the "master".

4.11   Increasing the Hit Rate by Adding an L2 Cache

In some cases a very small cache will not have a high enough probability of having the needed data to keep the processor fed at the required rate.

The next step has been to add another level of cache off the processor chip which is larger and runs at a speed between that of the first level cache and the main memory. At the lower speed, with a narrower bus and placed close to the processor, the ringing effect is reduced. Since the cache is now external it can be larger than an on-chip cache. It still needs to have measures taken to increase the probability that the data the processor wants next that is not in the first level cache is in the second level cache. This implies that an amount of logic will be added to the L2 SRAM to make it an effective cache.

It is also necessary to ensure that both the data in the cache and that in the main memory are updated as required so the latest data is available to the processor. This is called maintaining coherency between the cache and the main memory.

4.12   Operations to Ensure Cache Coherency

Various types of cache operation can be used to ensure coherency between the information in the cache and that in the main memory. Common cache coherency strategies are write-through and copy-back [1].

4.12.1   Write-Through

The write-through technique entails writing data simultaneously to the cache and to the main memory to ensure that the same data is available to both. There are several drawbacks to this method.

One drawback of this method is that the cache write access cycle is slowed down by the requirement of main memory access. It is possible to speed up the access by temporarily storing the data in a buffer which alleviates the wait for a main memory write at the expense of added complexity for the cache controller.

Another is that main system bus bandwidth is used with every access. Main memory access on every write can be avoided if the data is held in a buffer, but then the main memory and the cache are not coherent.

4.12.2 Copy-Back

Using a copy-back operation, the data written into the cache by the processor is not written into main memory until the time for that line of data to be replaced in the cache. This technique entails the use of a flag bit which can be set to indicate if that location in the cache has been written into. When the location has been written into and the flag bit is set, then the bit is said to be "dirty". The word at that location must be written into main memory before it is destroyed.

If the flag bit is not set, then the data at that location of cache is the same as at the corresponding address in main memory and the cache data can be destroyed. An example of a cache subsystem which illustrates these concepts follows.

4.13 External Cache Subsystems

An L2 cache subsystem normally consists of a controller, a cache tag SRAM and a cache data SRAM. A simple 40 MHz cache subsystem can be constructed with fast asynchronous SRAMs and fast logic as shown in the 128K Byte cache module in Figure 4.10.

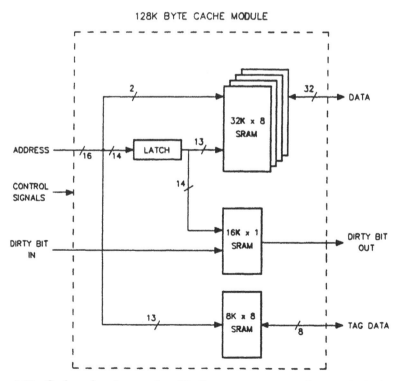

Figure 4.10 Cache subsystem made with discrete components (source: SGS-Thomson)

Here the data cache is composed of four 32K×8 SRAMs on a 32-bit bus. The tag is a single 8K×8 SRAM and the dirty bit is stored in a 16M×1 SRAM.

For high speed processors a faster cache subsystem with dedicated cache chips tends to be used. An example of a 66 MHz Pentium processor cache subsystem is shown in Figure 4.11.

The subsystem consists of a Pentium processor, a cache controller, a cache tag SRAM and a 256KB bank of 32K×18 data cache SRAMs. The address from the processor is checked by the cache tag SRAM at the beginning of each access. A match result is then delivered to the cache controller which in turn accesses the cache data SRAM bank if there is a match. Each of these dedicated cache SRAMs are described in more detail in the following sections.

4.13.1 Types of SRAMs Used for Cache Tags

SRAMs used for the tags can be small since only the address bits need to be stored in the tag. Since the length of the tag word depends on the size of the address space in the main memory, tag RAMs can be varying widths. For example, the 15 address bits required by the 32K×18 cache data RAM shown in the subsystem of Figure 4.11 can be stored in a tag RAM with 16-bit word length.

The depth of the tag RAM determines the number of addresses that can be stored. A block diagram of a 4K×18 tag RAM developed by Cypress [6], which could be used in the subsystem shown In Figure 4.11, is illustrated in the block diagram in Figure 4.12.

Each word in this tag RAM contains a 16-bit address tag field and a 2-bit status field. Up to 4096 main memory addresses can be stored in the 16-bit fields. The 12 address lines (indicating addresses in the tag, i.e. "sets" or "indices") select one of

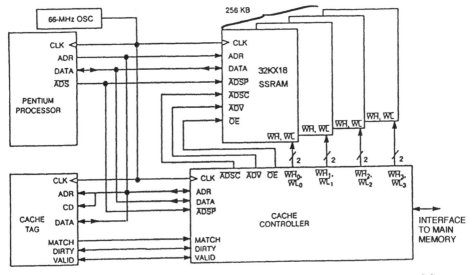

Figure 4.11   Pentium processor 256KB cache subsystem (source: Cypress [6])

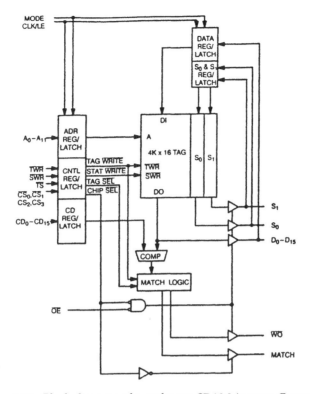

Figure 4.12   Block diagram of a cache tag SRAM (source: Cypress [6])

the 4096 16-bit words (main memory addresses) stored in the tag RAM. This 16-bit tag address is matched against data presented at the Compare Data inputs.

The matched outputs are compared against the two status bits in the tag RAM word. In multiprocessor operations the status bits can be used to store multiprocessing status information. In uniprocssor applications the two status bits are assigned as the valid bit and the dirty bits for implementing write-through or copy-back cache policies.

When the 16-bit tag address is matched against data presented by the processor at the compare data inputs, the match must be qualified by the valid bit of the chosen word. If the valid bit is set, the match is asserted.

An example of tag match timing is shown in the timing diagram in Figure 4.13.

The timing diagram shown illustrates a latch mode in which the LE input asynchronously samples the level of the other inputs. When LE is high and the mode input is high, the input latches are transparent and the inputs are allowed to flow into the tag. When LE is low, the latches are closed and the inputs can no longer flow in.

A tag compare cycle is initiated when tag select (TS\) is high. The tag entry is selected by A0–A11 and the 16-bit address stored in the tag is compared against the address requested by the processor which appears on CD0–CD15. The compare result is delivered to the match logic.

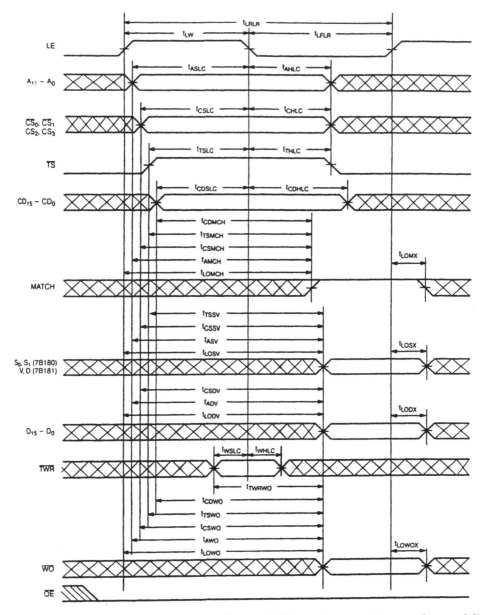

Figure 4.13 Tag match timing showing a hit (tag RAM a-sychronous) (source: Cypress [6])

The match output is driven high, indicating a match if the compare is successful or if the compare is successful and the bit is valid if the valid or dirty bit is set. The write output WO\ is also asserted when a match is detected. This signal may be connected directly to the write input of the cache data RAM or it may be connected through the cache controller as shown in the subsystem at the beginning of this section.

It is also possible to operate this tag RAM synchronously. In this case the mode input is low and the various inputs are sampled on the rising edge of the clock input (CLK) as shown in Figure 4.14.

Normally tag RAMs can be cascaded by connecting appropriate address lines to the chip select inputs. A RAM may be added for each chip select available.

By using dedicated tag RAMs, a wait state can be eliminated for the processor. This is due to the tag bit comparator on the cache tag SRAM that generates a "match" output to the controller much faster than the controller could provide the match off the RAM chip.

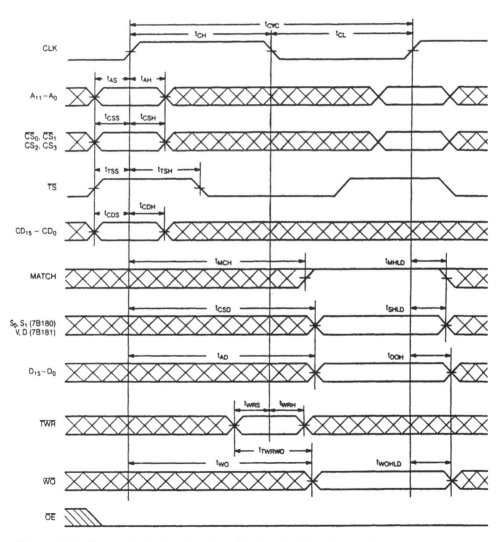

Figure 4.14 Tag match timing showing a hit (tag RAM sychronous) (source: Cypress [6])

4.14   SRAMs Tailored for Cache Data RAMs

4.14.1   *Asynchronous Cache SRAMs*

As long as processors did not require speeds over about 33 MHz for memory access, it was possible to build an SRAM cache which did not have wait states using two banks of interleaved asynchronous SRAMs. This solution is expensive due to the two banks of SRAMs required and it also consumed more power than a single bank of SRAMs [9].

Over 33 MHz the difficulties using standard asynchronous SRAMs for data caches became significant and the industry turned to single bank caches composed of synchronous SRAMs with burst mode features to accommodate the burst mode capability of many of the modern higher speed processors.

4.14.2   *Synchronous Cache SRAMs*

The 256KB bank of data SRAMs shown in the subsystem in Figure 4.11 was composed of four synchronous 32K×18 cache SRAMs. Similar synchronous cache SRAMs are widely available in 32K×9, 32K×18, and 32K×32 or 64K×18 configurations.

The 256KB cache subsystem shown in Figure 4.11 could be simply upgraded to a 512KB subsystem by substituting 64K×18 cache SRAMs for the 32K×18 cache SRAMs and adding an address line. Such systems are frequently built with the extra address line in preparation for such an upgrade when the memory technology is available in the higher density.

A block diagram of a 32K×18 synchronous cache SRAM with input registers is shown in Figure 4.15. Synchronous SRAMs have become widely sourced as cache data RAMs for the 586 class of processors to which they can interface with minimum glue logic. They have synchronous self-timed write and an asynchronous output enable [6].

To write to this part in a cache subsystem, the "processor address strobe input" (ADSP\) on the SSRAM is sampled by the processor at the rising edge of the clock (CLK). If this input is active the address request from the processor (A0–A15) is latched into the input register on the cache data RAM as shown in the timing diagram in Figure 4.16(a).

On the next clock the addresses flow on into the RAM. The one-cycle delay is to permit the cache tag logic to use this clock period to perform address comparisons or protection checks.

If the write is allowed to proceed, the write input to the cache data RAM will be pulled low by the cache controller logic before the next rising clock edge. This permits the data at DQ0–DQ15 to be written into the cache data RAM.

A write cycle can also be initiated by the controller using the "controller address strobe input" (ADSC\) in which case the data is written on the rising edge of the clock in the first cycle in which ADSC\ is active. This is shown in the timing diagram in Figure 4.16(b).

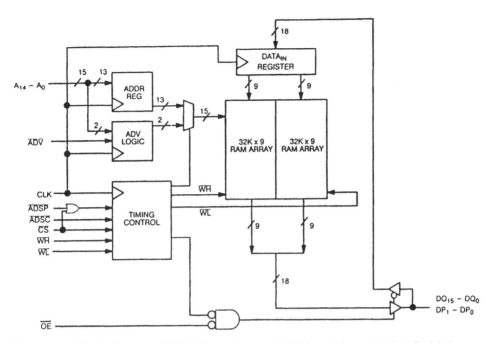

Figure 4.15   Block diagram of 32K×18 synchronous SRAM used for cache data RAMs (source: Cypress [6])

4.14.3   Burst Mode on Synchronous Cache SRAMs

These synchronous cache data SRAMs are equipped with burst mode capability. In burst mode a two-bit on-chip counter captures the first address in a burst and increments the address automatically for the rest of the four-bit burst access.

Two types of burst are possible–interleaved (such as required by the Intel type processors) and linear for processors using a linear burst sequence (such as the Motorola 68×××series). An example of the counter implementation for an interleaved burst sequence is shown in Figure 4.17(a) and for a linear burst sequence in Figure 4.17(b) [6].

An example of a timing diagram for a burst read is shown in Figure 4.18 [6].

The addresses are sampled on the rising edge of the clock (CLK). The asynchronous output enable is active (OE\ is low). At a time t_{CDV} after the rising edge of the clock the first data bit appears on the output.

The address is incremented as the rising edge of the next clock samples ADV\, which is active, and also the output turns off at a time t_{DOH} after the rising clock edge, and the data is available at the outputs after 8.5 ns.

To equal the speed of the burst SSRAM, two banks of 15 ns asynchronous SRAMS would need to be interleaved thereby nearly doubling the cost of the system.

An additional advantage of using the burst SSRAMs to replace two banks of interleaved asynchronous SRAMs is that a 9 ns burst SSRAM has a two cycle latency for read at 66 MHz, whereas two interleaved 15ns standard asynchronous SRAMs have a three-cycle latency. This is illustrated in the timing diagrams in Figure 4.19 [9].

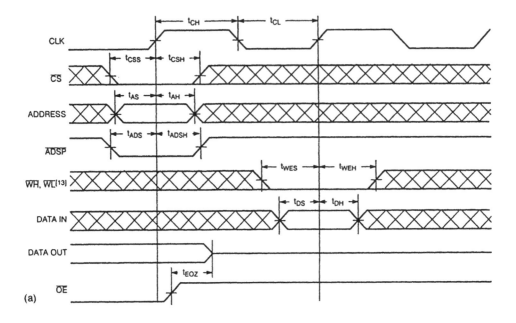

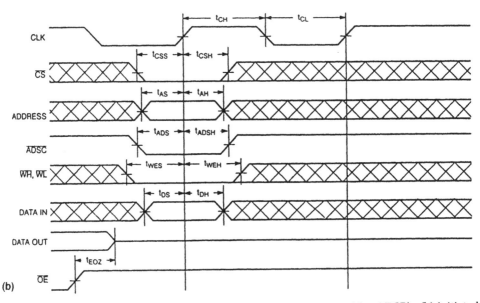

Figure 4.16 Synchronous cache data RAM write timing: (a) initiated by ADSP\; (b) initiated by ADSC\ (source: Cypress [6])

Figure 4.19(a) shows the timing diagram for two banks of interleaved 15 ns access time asynchronous SRAMs running in a 586 class processor system with a 66 MHz clock. One cycle is consumed by the address set-up time of the processor. The address is latched into the cache SRAM on the rising edge of the second clock and

First address	Second address	Third address	Fourth address
A_{x+1}, A_x	A_{x+1}, A_x	A_{x+1}, A_x	A_{x+1}, A_x
00	01	10	11
01	00	11	10
10	11	00	01
11	10	01	00

(a)

First address	Second address	Third address	Fourth address
A_{x+1}, A_x	A_{x+1}, A_x	A_{x+1}, A_x	A_{x+1}, A_x
00	01	10	11
01	10	11	00
10	11	00	01
11	00	01	10

(b)

Figure 4.17 Counter implementation for burst sequences: (a) interleaved; (b) linear (source: Cypress [6])

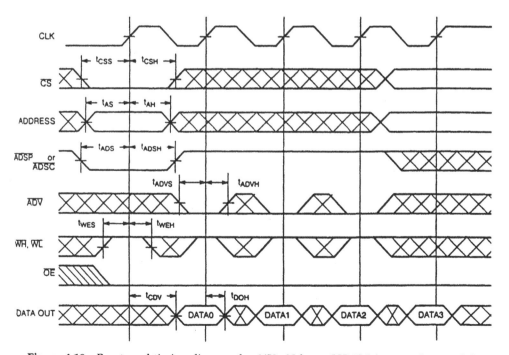

Figure 4.18 Burst read timing diagram for 64K×18 burst SSRAM (source: Cypress [6])

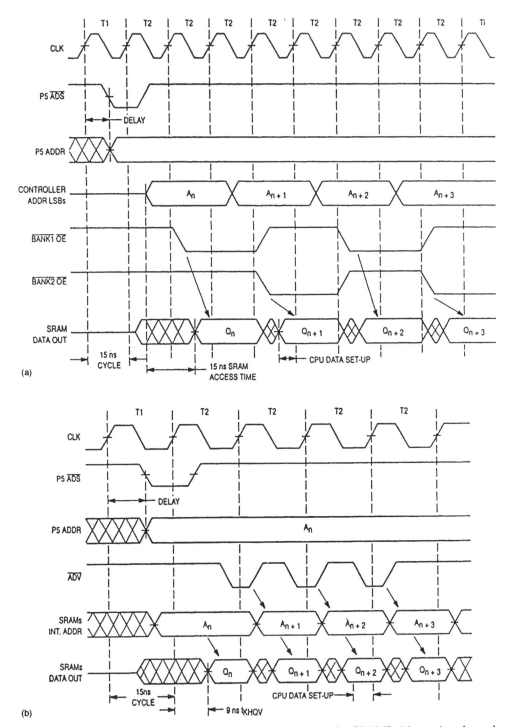

Figure 4.19   Comparison of the burst read cycles for two cache SRAMS: (a) two interleaved banks of asynchronous SRAMS (3–2–2–2) burst read; (b) singe bank of burst SSRAMS (2–1–1–1) burst read (source: Motorola [9])

the data (main memory addresses) are available on the third clock and every other cycle thereafter. The timing in terms of the clock is therefore 3-2-2-2 best case. If a large cache is used, the loading can increase the delay still further [10].

A single asynchronous SRAM bank with 7 ns access time would be required for a 66 MHz clock to eliminate wait states. This includes the 5 ns processor address set up time plus the 7 ns access time plus two 1.5 ns transition times. Because of the added delay of the tag bit comparison, zero wait state performance is generally not possible. The faster part would also be more expensive than the parts in the 15 ns SRAM technology.

The burst SSRAM has the advantage that a part in the less expensive 15 ns technology with the synchronous architecture can achieve a 9 ns burst. Figure 4.19(b) shows the timing diagram of a single bank of burst SSRAMs with a 66 MHz clock. The address is set up in the first cycle, then 9 ns into the second cycle the data out is valid. Data is then available during the burst on each rising edge of the clock. The timing in terms of the clock is therefore 2-1-1-1.

Burst SSRAMs are available in the various upgrade configurations shown in the TQFP package pinouts in Figure 4.20. These include 32K×9, 32K×18, 64K×18 and 32K×32(36).

A die photo of a 64K×18 synchronous cache SRAM was shown in Figure 1.3. This part was manufactured in 0.5m full CMOS and had a typical clock to data output time of 6.5 ns.

4.14.4   Pipelined Burst SSRAMs

Burst synchronous SRAMs are also available with a pipelined architecture for higher speed systems. The pipelined architecture is implemented with registers on both the data inputs and data outputs of the SRAM. While the burst SSRAMs discussed in the previous section had asynchronous data output, the pipelined SSRAMs are fully synchronous.

The advantage of a pipelined architecture is that the cycle time can be minimized by equalizing the delays through each segment of the pipeline.

The block diagrams in Figure 4.21 compare a 32K×32 pipelined synchronous burst SRAM with input and output registers with a "flowthrough" burst SSRAM with only input registers.

Read timing diagrams illustrating the differences in operation of a pipelined and and flowthrough 32K×32 SSRAM are shown in Figure 4.22.

The initial access of the pipelined burst SSRAM is one cycle longer than that of the flowthrough part due to the additional output register. However, subsequent reads in the burst are faster. For the SSRAMs shown in Figures 4.21 and 4.22, the access time of the flowthrough SRAM is specified at 10 ns and the corresponding pipelined SRAM is specified at 5 ns.

A write pass-through feature on the pipelined part makes data which has been written immediately available at the output register during the read cycle following a write. This is illustrated in the read/write timing diagram in Figure 4.23.

The part also incorporates a pipelined enable register to allow depth expansion without penalizing system performance.

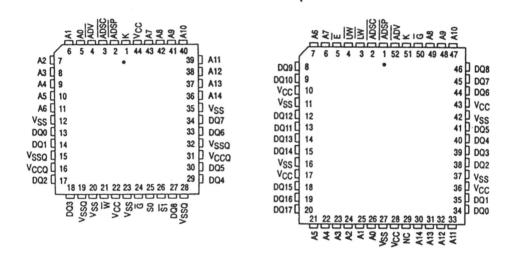

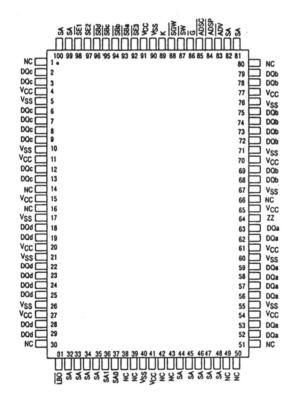

Figure 4.20   Burst SSRAM pinouts for 32K×9, 32K×18 and 32K×32 (source: Motorola [16])

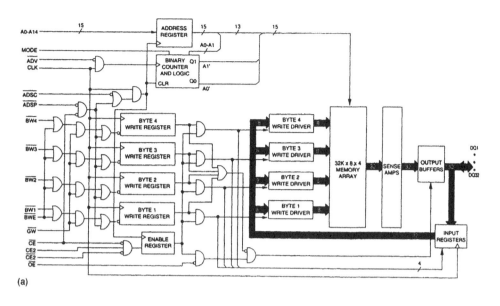

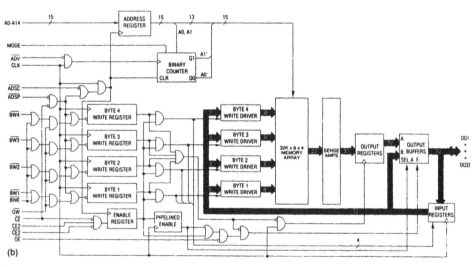

Figure 4.21   Block diagrams of 32K×32 burst SSRAM: (a) flowthrough (no output registers); (b) pipelined (with output registers) (source: Micron Technology)

4.15   Use of Parity in Caches

One topic that has been ignored up to this point is parity. Memories that have ×9 (or multiples of ×9) instead of ×8 I/O's have an extra bit to check the parity of the byte.

Since the prevalent cache SRAMs at the 256K and 512K density had a parity bit, one would expect that a cache that used a 32K×9 or 32K×18 type of SRAM would be upgraded to a 32K×36 SRAM. While this has happened in the high performance

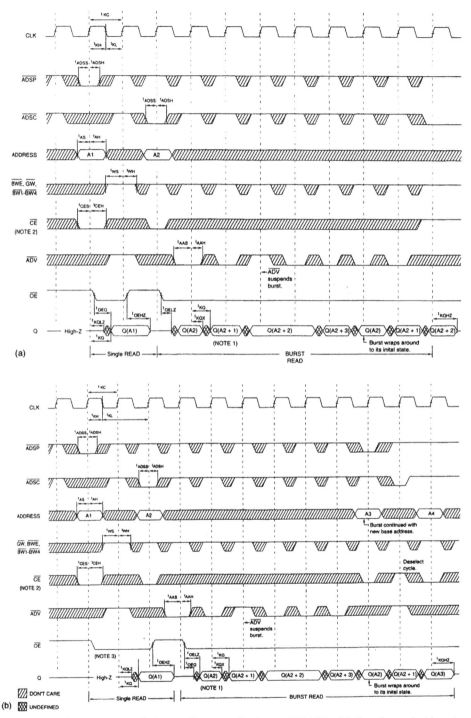

Figure 4.22 Comparing read timing diagrams for burst SSRAMs: (a) flowthrough (showing two-cycle latency); (b) pipelined (showing three-cycle latency but faster burst) source: Micron Technology [12])

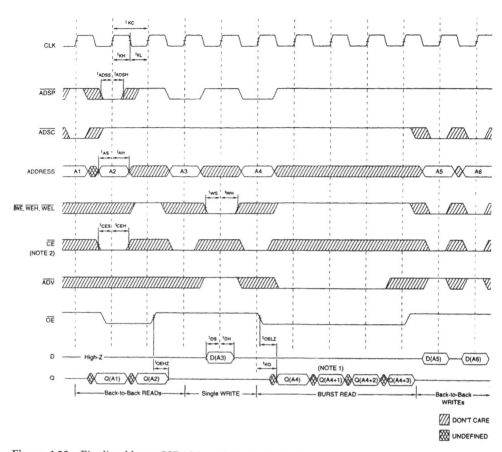

Figure 4.23 Pipelined burst SSRAM read/write timing diagram (source: Micron Technology)

workstation area, PCs are upgrading to 32K×32 and perhaps 64K×16 Burst SSRAMs which eliminate the cost of the parity bits.

Parity is being eliminated for most PC applications for several reasons. The first is that the rate of errors detected by parity is very low. The second is the futility of parity detection in PC applications. In most cases the only result of a detection is that the system locks up and must be rebooted. This can be frustrating but it is infrequent.

The third is that newer PC operating systems do not flag the parity error the way the older DOS operating system did. With DOS, the parity error message appeared on the screen so the user was aware of the occurrence. Newer operating systems can also lock up, but since a lockup can be caused for other reasons as well, the perception of parity as a problem is missing in the average user. The user will therefore be reluctant to pay additional amounts for a system with parity detection.

In summary the reasons for eliminating parity are the low rate of errors detected by parity in most PC applications, the fact that parity errors are not flagged for users in the newer PC operating systems, and the fact that the average user is not willing to pay for unperceived problems.

In workstations, however, parity remains. This is because the larger size of the memories in workstations causes the potential for errors detectable by parity to increase. In addition many types of applications run on workstations tend to be more sensitive to errors so the requirement for parity by the user is greater.

It should be expected, therefore, that the deepest cache SRAMs which would tend to be used in the largest caches would continue to carry the parity bit. For example, the 32K×32 burst SSRAM which tends to be used in 256KB caches in PCs has dropped the parity bit, but the 64K×18 burst SSRAM which is used in 512KB caches in workstations has retained it. As PCs migrate to the 512KB cache, we might expect to see the parity bit dropped.

Bibliography

1. Handy, J., *The Cache Memory Book*, Academic Press, 1993.
2. Przybylski, S. A., *Cache and Memory Hierarchy and Design*, Morgan Kaufmann, 1990.
3. Jouppi, N. P., Introduction to computer architecture, Notes from ISSCC Short Course, 1993.
4. Hennessy, J. L. and Patterson, D. A., *Computer Architecture: a Quantitative Approach*, Morgan Kaufmann, 1990.
5. *NEC Static RAM Data Book*.
6. *Cypress High Performance Data Book*, 1994.
7. *Integrated Device Technology Databook*.
8. Przybylski, S. A., Memory hierarchy design", Notes from ISSCC Short Course, February 1993.
9. Peters, M., A zero wait state secondary cache for Intel's Pentium, *Motorola Fast SRAM Databook*, 1994.
10. Ohr, S., Fast cache designs keep the RISC monster fed", *Computer Design*, January 1995.
11. Prince, B., *Semiconductor Memories*, 2nd Edition, John Wiley and Sons, 1992.
12. *1995 SRAM Databook*, Micron Semiconductor.
13. Smith, A. J., CPU Cache Memories, Tutorial at Hot Chips Symposium, August 1993.
14. Tuite, D., Cache architectures under pressure to match CPU performance, *Computer Design*, March 1993, 91.
15. Hochstedler, C., Cost-effective PC cache memory, *Computer Design*, March 1993, 93.
16. *3Q95 Fast Static RAM Component and Module Databook*, Motorola.
17. Peters, M., Private Correspondence 1995.
18. *High Performance Static RAMs*, I.D.T. 1994
19. *High Density Cache SRAM Report*, Memory Strategies International, 1995.

5 Evolution of Fast Asynchronous DRAMs

5.1 Overview

Dynamic RAMs are used predominantly in main memory in most computer systems today. They are the cheapest form of random access semiconductor memory with an ongoing specification trade-off between speed and cost. While high performance computer systems frequently have a faster cache of more expensive SRAM between them and the processor to reduce the penalty for the DRAM speed somewhat, the penalty for a cache miss still needs to be dealt with by the DRAM.

The major benefit of a DRAM is that it is a low cost commodity product. As such it has taken market from every other random access data storage that has appeared. Figure 5.1 shows the market share of the DRAM as a percentage of the total memory market over time.

With the development of faster processors, the speed demands on the system memory have increased. Adding more cache SRAM is expensive, so there has been a continual effort on the part of DRAM manufacturers to increase the speed of the part while maintaining the low cost expected of the DRAM.

If we consider the basic read cycle of the DRAM, there is a small increase in speed due to the technology over time but it is not enough to match that required by the processor. Table 5.1 gives an indication of the increase in speed of the read cycle time and datarate of a ×1 organized DRAM by density. The approximate year of first production of each density DRAM is indicated.

Processor internal speed capability over time is plotted in Figure 5.2 compared to that of a DRAM read access cycle and also a DRAM page cycle which was the most common DRAM access cycle actually in use up to 1995. While the DRAM speed has increased over time, it has not kept up with the speed of the microprocessor. The microprocessors, as well as enjoying the gains due to technology, have also improved their cycle times by employing a mixture of process and architectural techniques such as pipelining, parallel executions, etc. It should be noted that the external bus of the

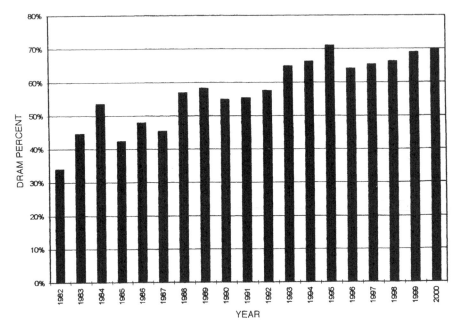

Figure 5.1   DRAM share of the total memory market, 1982–2000 (source: Dataquest, Electronics Survey, MSI)

Table 5.1   Trends in speed of DRAM read cycle by density

Density (bits)	Year	T_{CYC} (ns)	Speed (MHz)	Datarate of Mb/sec	×1 DRAM MB/sec
16K	1978	400	2.5	2.5	0.3
64K	1981	270	3.7	3.7	0.5
256K	1984	160	6.2	6.2	0.8
1M	1987	140	7.1	7.1	0.9
4M	1990	120	8.3	8.3	1.0
16M	1993	110	9.0	9.0	1.1
64M	1996	100	10.0	10.0	1.2

processor rarely runs at the internal speed capability due to the speed mismatch with the DRAM.

Table 5.1 also indicates the datarate of the DRAM which is the rate at which data can be obtained from the memory. The real concern from a system standpoint is that sufficient data be available from the memory when it is required by the processor to avoid wait states.

If, for example, data could be obtained 100 bits at a time, then the DRAM would only need to be 0.01 times as fast. Datarate is, therefore, an important concept which has given both processors and DRAMs some flexibility in dealing with the speed

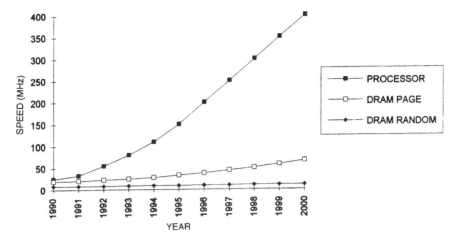

Figure 5.2   Processor and DRAM speed trends (MHZ)

mismatch problem between them. Datarate is usually calculated in megabytes per second (MB/sec) for historical reasons, where a byte is equal to eight bits.

A related problem is that of granularity which is the minimum increment of memory required by the system. The earliest DRAMs were organized ×1. This meant that one bit of data could be obtained on every access. If a system had a granularity of 16KB which was a reasonable amount of main memory in this time period, then eight 16K DRAMs placed side by side on an 8-bit bus could be used. At every access all eight DRAMs were accessed and eight bits of data appeared on the bus. A read cycle took about 400 ns which could support a 2.5 MHz processor with 2.5MB/sec of memory.

When, however, the industry moved to the 64K DRAM, only two 64K×1 DRAMs were needed to provide the 16KB of memory. However, to maintain at least the 2.5MB/sec datarate required by the system with a 2-bit bus the parts would have had to run at 10 MHz (2 parts × 10 MHz × 1 byte/8 bits).

The 64K DRAM ran, however, at only 3.7 MHz, so at least six of the ×1 organized parts on the bus were still needed. Rather than redesign the system for a 6-bit bus, eight 64K×1 DRAMs were used. This 64KB of memory was greater than the minimum 16KB required granularity of the system and meant that the user was paying for more memory than was needed.

Another option used was to go to a 4-bit wide architecture on the DRAM. Two 16K×4 DRAMs on the 8-bit bus gave a 16K×8 system which maintained both the granularity and the bus width. The datarate even increased slightly since the 64K DRAMs, even with a wider organization, were slightly faster than the 16K DRAMs. This trade-off between speed, granularity, and bus width will be seen to play a factor in the efforts to improve the datarate of the DRAM.

All of the fast DRAM techniques which follow are attempts to deal with these basic trade-offs and maintain the cost-effective commodity nature of the DRAM which is ultimately its only justification for existence.

5.2   Basic DRAM Operation

As the low cost commodity RAM used in main memory, many trade-offs have been made in the past to arrive at the present DRAM. First let us look at how the DRAM became so much slower than its cousin, the faster but more expensive SRAM.

The DRAM evolved as a cheap alternative to the SRAM. It paid for its reduced chip size with increased circuit and cell complexity and slower speed. The large stable six-transistor cell of the SRAM was replaced by a smaller cell consisting of a large leaky storage capacitor and an access transistor which permitted a smaller and lower cost chip. A penalty of the high capacitance was slower speed. Another penalty was that the charge on the capacitor needs periodic refresh so refresh clocks and counters need to be supplied and time must be taken for refresh.

The addresses also were time-multiplexed to halve the number of address pins. This reduced the package size and cost but added both to the time required to access the data and to the complexity of the DRAM.

A block diagram of a ×1 organized DRAM is shown in Figure 5.3. The address inputs are multiplexed into separate row and column address buffers. The row and column addresses are decoded and strobed sequentially by pulses from the RAS and CAS clock generators. The refresh addresses and refresh counter are provided on the chip and periodically are required to cycle through the cells.

Each sense amplifier both reads and restores the data to the cell. Since an access bleeds the charge off of the cell capacitor, it destroys the bit stored there and must restore it before ending the cycle. The data from an entire row (or wordline) is held on the sense amplifiers.

The column decoder selects the bit or bits that are addressed. In a read cycle these bits are fed onto the datapath to the output buffer of the DRAM. In a write cycle the data feed into the data-in buffer and into the sense amplifiers which are selected by the column decoder.

A timing diagram can be used to examine the delays inherent in the DRAM. A typical read timing diagram for a 4M×1 DRAM corresponding to the block diagram shown above is shown in Figure 5.4.

Operationally 22 address bits are required to decode 1 of 4Mb storage locations. First the 10 row address bits are set up on the address pins and held until they stabilize t_{ASR}. They are then latched into the row address buffer by the falling edge of the RAS\ pin which also activates the sense amplifiers. The addresses must be held on the inputs until RAS\ stabilizes t_{RAH} [9].

The selected row (wordline) out of 1024 rows is driven high and the data in the 4096 memory array cells on that wordline is fed out onto the bit lines (columns) and latched into the 4096 sense amplifiers.

The 12 column address bits are set up, allowed to stabilize t_{ASC}, and latched into the column address buffer by the falling edge of the CAS\ pin which also turns on the output buffer. The CAS\ pin must be held low until it stabilizes (t_{CAH}).

The column address decoder selects the one bit out of 4096 that is addressed and feeds it to the data output buffer.

It should be noted that the column address buffers are also activated on the falling edge of RAS\ and act as flowthrough latches as long as CAS\ remains high. This

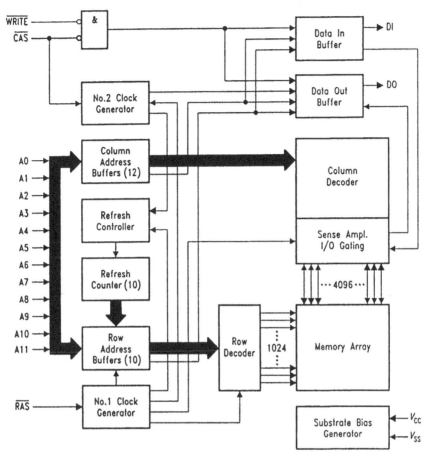

Figure 5.3 Block diagram of a 4M×1 DRAM (source: adapted from Siemens 4 M DRAM document)

permits data retrieval to begin as soon as the column address is valid rather than waiting until the CAS\ pin falls. (The column address set-up must wait a minimum time ($t_{RAD} > t_{RAH}$) after the RAS\ pin falls to permit the row addresses to stabilize.) This was an early improvement in the DRAM timing and is an example of pipelining for speed in the DRAM architecture.

The delays inherent in setting up the read access cycle are the minimum timings given on the datasheet. These include the row address set-up time (t_{ASR}), transition times (t_T), row address hold time (t_{RAH}), row to column delay time (t_{RCD}), column address set-up time (t_{ASC}), read command set-up time (t_{RCS}) and column address hold time (t_{CAH}). In addition, prior to the next read access there is a delay for CAS precharge (t_{CRP}) and RAS precharge (t_{RP}) which adds to the cycle time.

The actual data is expected out of the DRAM at a time specified by the maximum values of the RAS\ access time (t_{RAC}), the CAS\ access time (t_{CAC}), the address access time (t_{AA}) and, if output enable is used, the output enable access time (t_{OEA}). These parameters all depend on the internal speed of the various required DRAM operations and are not too amenable to improvement.

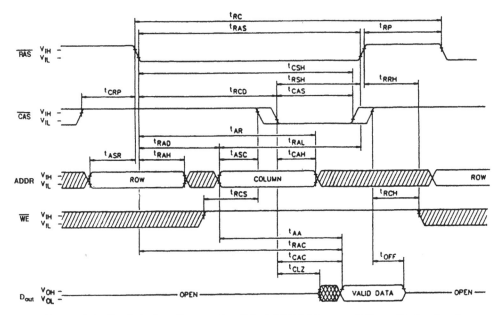

Figure 5.4 Read timing diagram for 4M×1 DRAM (source: Micron Technology)

Most effort in DRAM speed improvement has therefore focused either on getting more bits out of the DRAM on one access given these constraints, on pipelining the various required operations in such a way as to minimize the total time, or in segmenting the data in such a way that some of the operations are eliminated for a given set of accesses. These three techniques are examined in the following sections.

The write operation is similar to the read operation except that after the RAS hold time is satisfied, the WE\ pin is brought low and allowed to stabilize for a time t_{WCS} and valid data is presented at the I/O pins and allowed to stabilize for a time t_{DS} before being clocked in by the CAS\ pin going low.

The data is held on the inputs for a time t_{DH} while the data is latched into the sense amplifiers selected by the column decoder. It is then latched into the corresponding location in the memory array during the restore operation.

Figure 5.5 shows an early write cycle for the 4M×1 DRAM.

5.3   Early Speed Improvements

The most obvious way to speed up the datarate is by usefully accessing more of the information contained in the array without having to go through the entire access cycle again. For example, if four bits of data could be obtained on one access, then the datarate would be quadrupled.

Two early methods were developed for obtaining multiple bits of data on one access. The first was to keep the one-bit width of the I/O port but take four bits of

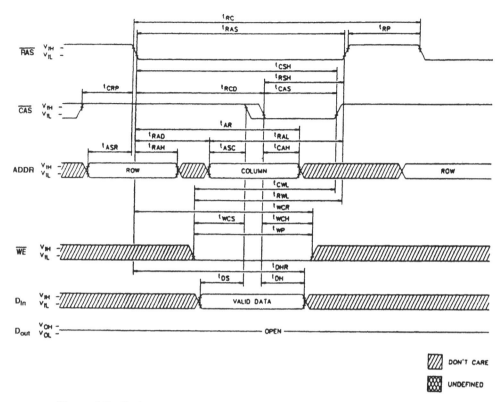

Figure 5.5 Early write cycle timing diagram (source: Micron Technology)

data out serially on every access; the other was to make the output buffer more than one bit wide. Both methods were used.

5.3.1 Nibble Mode

The method of clocking four rapid bits of data out serially on each access is called nibble mode [9]. An example of a nibble mode read timing diagram for a 4M×1 DRAM is shown in Figure 5.6 and of a nibble mode write timing diagram in Figure 5.7.

In a nibble mode read or write access, the first bit is accessed in the usual way by setting up the row and column addresses and latching them in with the RAS\ and CAS\ pins. The next three sequential bits can be accessed by cycling CAS\ while RAS\ remains low without the need for additional external row or column addresses.

The cycle time for the last three bits is considerably shorter than that of the first bit since all of the external address set-up and hold times are eliminated.

The nibble mode cycle time is CAS\ precharge time (t_{PC}) + CAS\ pulse width (t_{CAS}) + twice the transition time ($2t_T$). When calculating the total nibble mode cycle

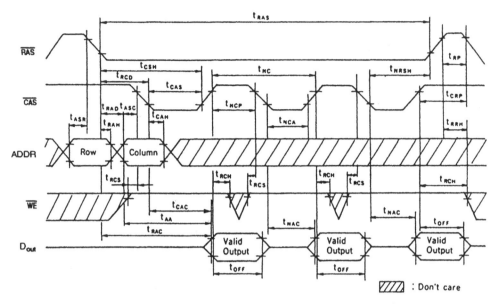

Figure 5.6　Nibble mode read cycle time (source: Hitachi)

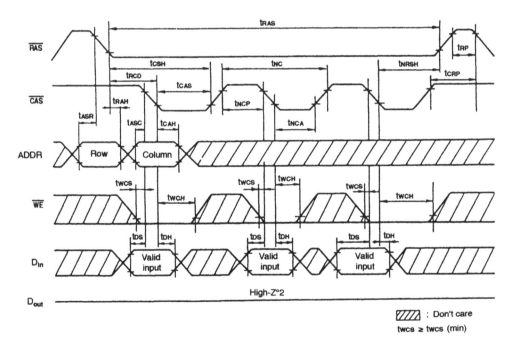

Figure 5.7　Nibble mode write cycle time (source: Hitachi)

time in a real system, it is also necessary to average in the longer cycle time required to obtain the first bit.

A schematic block diagram for a 4M×1 DRAM with nibble mode addressing is shown in Figure 5.8.

A nibble selector on the column decoder selects the three bits which follow the bit which was addressed.

The first bit of the nibble selected is determined by the row and column addresses supplied. Thereafter, the least significant bit of the row address and the most significant bit of the column address provide binary combinations for the next three addresses.

In the case of the 4M×1 DRAM used in the example, row A10 (the least significant bit) and column A10 (the most significant bit) provide the binary bits for initial address selection. For the next three accesses the falling edge of CAS selects an address formed from the next bit of the binary sequence

$$(0,0) \rightarrow (0,1) \rightarrow (1,0) \rightarrow (1,1)$$

which wraps around as shown. It is possible to start on any address of the nibble sequence.

Since the four bits were fed out serially through the one-bit output port, nibble mode did not change the pinout and thus did not increase the size of the DRAM package.

Nibble mode is no longer a significant factor in the DRAM market. It is, however, a first example of a burst mode which is just becoming a factor in modern DRAMs. The modern burst modes will be discussed more in a later section.

5.3.2 Wide I/O

It is also possible to work with more data bits on a given access by making the input/ output (I/O) port of the DRAM wider. Going from one data bit to multiple data bits entails dividing the array into segments which are simultaneously accessed feeding

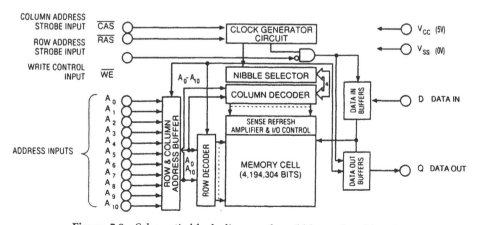

Figure 5.8 Schematic block diagram for nibble mode addressing

multiple banks of sense amplifiers. This is shown in the functional block diagram for a 1M×16 DRAM in Figure 5.9.

The 1M×16 array, in this figure, is divided into 16 arrays of 1024 × 1024 bits. Each array has 1024 sense amplifiers attached to its 1024 columns. A 1K cycle refresh is assumed in this example.

During a read cycle 16 banks of 1024 sense amplifiers are active. After the RAS\ clocks in the row address, each sense amplifier bank contains data from the addressed 1024-bit row in its array. When the column decoder then selects the addressed column, the bits from this column in all 16 banks are moved over parallel internal data paths between the sense amplifiers and the I/O port, avoiding the time and additional buffering required for a serial access stream. The time required is about the same as for obtaining just one bit.

The datarate gained is significant. A 100 ns cycle time DRAM goes from a datarate of 10M-bit per second for the ×1 part, to a datarate of 80M-bit per second for the ×8 and 160Mb per second for the ×16.

There are, however, several trade-offs involved in going from a single to multiple I/O ports on a DRAM.

The first is cost. The size of the DRAM chip, and potentially that of the package, tends to increase as the I/O widens because of both the additional I/O pins and the additional silicon area used in segmenting the array. This is illustrated in Figure 5.10 which shows the pinouts for different widths of the 4M DRAM for the SOJ package. This cost adder is partially compensated for by a decrease in test time for wider DRAMs.

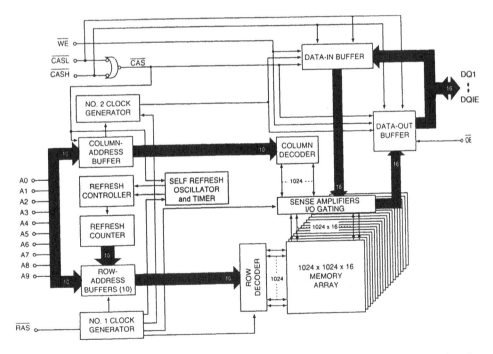

Figure 5.9   Functional block diagram for a 1M×16 DRAM (source: Micron Technology)

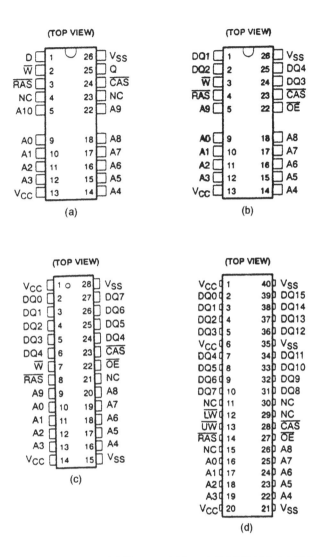

Figure 5.10 Pinout comparisons of different widths of 4M DRAM: (a) 4M×1 ; (b) 1M×4;
(c) 512K×8; (d) 256K×16 (source: Hitachi)

There is no difference in package size in going from the ×1 to the ×4, for both are
in a 22-pin package. The separate input and outputs on the ×1 were changed to
common I/O in the ×4 so only two I/O pins needed to be added rather than six. An
output enable was also added, but there is one less address. The net gain is therefore
two pins.

So that the ×1 and ×4 could be made on a common chip, the package was kept the
same. There is clearly little cost differential between these two parts.

The ×8, however, added another four I/O pins while losing one address and
adding a "no connect" (NC), so the package grew by four pins. The ×16 added
another 8 I/O pins, losing only one address pin in the process. An extra control

for the upper and lower bytes was also added. The result was a move to a more expensive 40-pin package.

Another drawback of wider I/O's is that multiple outputs switching simultaneously can draw enough current to set up ringing in the ground circuit, a phenomenon called ground bounce, and the data on the output can not be read until it has stabilized. This can result in parts with wider I/O being slower than parts with narrow I/O. This factor limited the implementation of outputs wider than four bits for a number of years. There is further discussion of the ground bounce phenomenon in Chapter 3 Section 7.

5.4   Special Access Modes

Another direction for DRAM speed enhancement has been segmenting the data in such a way that some of the necessary internal DRAM operations are eliminated for a given set of accesses.

This effort has been primarily directed at obtaining rapid access to the large amount of information held on the sense amplifiers during a normal read access. In this case the sense amplifiers are being used very much like the caches that were described in the previous chapter.

Recall that, during a normal access of a DRAM, after the RAS\ latches in the row address, a large number of bits are held on the sense amplifiers waiting for the column decoder to select the particular bits addressed. Figure 5.3, for example, showed 4096 sense amplifiers that are turned on during the read or write cycle for a 4M×1 DRAM and Figure 5.9 showed 16 banks each with 1024 sense amplifiers containing data during the read access of a 1M×16 DRAM.

Many techniques have been developed for rapid access of the data held on the sense amplifiers. Some of the more common of these are described in the following sections on page mode and fast page mode.

5.4.1   *Page Mode*

Probably the most commonly used method for rapid random access of the bank or "page" of data on the sense amplifiers is called *page mode*. In page mode a normal access is initiated by latching in the row address. The row address strobe is then held active, which maintains a page of data on the sense amplifiers. New column addresses are then repeatedly clocked in by the CAS\ thereby permitting rapid random access of the page of data. Page mode is faster than pure random access since it eliminates the row address set-up and hold times and the row precharge time.

The drawback of page mode is that the data must be segmented so that the probability is high that the desired data is on the page currently being accessed. If data occurs randomly throughout the memory array then page mode will be of little use. Many applications, particularly in the graphics area, are very suited to page mode access.

An improvement was quickly made on the original page mode leading to the "enhanced page mode" or "fast page mode" which is used on all modern DRAMs having this feature.

5.4.2 Fast (Enhanced) Page Mode

Fast, or Enhanced, Page Mode is the most important mode in DRAMs today. The difference between enhanced/fast page and the older page mode is that the older version initiated column address access on the falling edge of the CAS\ strobe so the column address set-up time was included in the page cycle time.

In the "enhanced" or "fast" version, the column address buffers are activated on the falling edge of the RAS\ strobe and remain open, acting as transparent latches when CAS\ is high, and latch the column address on the falling edge of CAS\. This means that address set-up begins as soon as the column address is valid rather than waiting until CAS\ transitions low.

A fast page mode read timing diagram is shown in Figure 5.11 and an early write page mode timing diagram is shown in Figure 5.12.

The page mode read sequence involved is as follows. First a normal RAS\ and CAS\ access is done to access the first data bit. RAS\ is then left low (on). When CAS\ goes high at the end of this first access, it turns the output buffer of the DRAM off (after a time t_{OFF}) thereby ending the first access. A new column address can now become valid and, after the CAS\ precharge time is met, can be latched into the column decoders with the falling edge of CAS\.

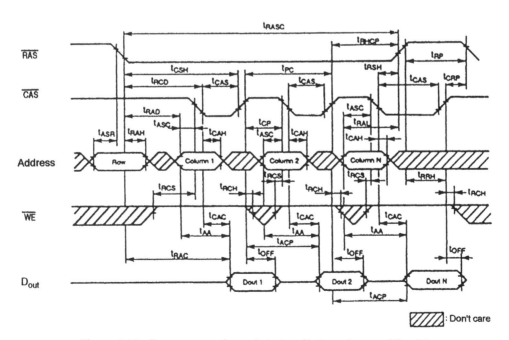

Figure 5.11 Fast page mode read timing diagram (source: Hitachi)

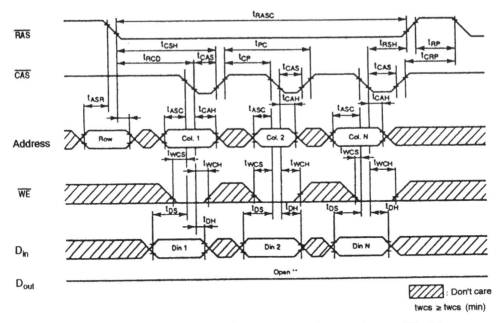

Figure 5.12 Fast page mode early write timing diagram (source: Hitachi)

The data is selected from the page held on the sense amplifiers and fed out onto the data path where it is available on the output buffer after time t_{CAC}. The falling edge of CAS\ also takes the output buffer from high to low impedance in preparation for the new data. After the data appears on the output, stabilizes and is read, the CAS\ goes high again, turning the output buffer off and allowing the next CAS\ cycle to begin.

The timing delay for fast page mode for the first access is the normal access time. Minimum timing delay for subsequent CAS\ accesses is determined by whichever one of the internal access timings – CAS\ access time (t_{CAC}), address access time (t_{AA}), or CAS\ precharge access time (t_{CPA}) – happens to be limiting for the particular part and timing set-up used.

In fast (enhanced) page mode, the critical parameter is the page cycle time (t_{PC}) which must obey

$$t_{PC} = t_{CP} + t_{CAS} + 2t_T > t_{CPA} + t_T$$
$$> t_{AA} + 2t_T$$

Consider, for example, a 5 V -6 part. Typical numbers are:

maximum delays: $t_{AA} = 30$ ns, $t_{CAC} = 15$ ns, $t_{CPA} = 35$ ns
minimum delays: $t_{CYC} = 110$ ns, $t_{CAH} = 10$ ns, $t_{CP} = 10$ ns, $t_T = 5$ ns
$t_{OFF} = 0$, $t_{CAS} = 15$ ns.

If we just add the minimum timings, we get a page cycle time,

$$t_{PC(min)} = t_{CAS(min)} + t_{CP(min)} + 2t_T$$
$$= 15 \text{ ns} + 10 \text{ ns} + (2 \times 5 \text{ ns}) = 35 \text{ ns}$$

Most data sheets will specify this as the minimum page cycle time. This value is normally not attainable, however, due to the other limitations on the page cycle time.

In this example, the limiting parameter on the cycle time is:

$$t_{PC} > t_{CPA} + t_T = 40 \text{ ns}$$

This minimum page cycle time is obtained when the addresses are valid before the CAS\ precharge begins, so that the maximum time is given for column address set-up. The shortest page mode cycle time is, therefore, that with maximum address set-up time.

An illustration for fast page mode with maximum address set-up time is shown in the timing diagram in Figure 5.13. In this case,

$$t_{PC(min)} = t_{CP} + t_{CAS} + 2t_T = t_{CPA} + t_T$$
$$= 10 + 20 \quad + 10 = 35 \quad + 5 = 40 \text{ ns}$$

Where t_{CAS} is not at its minimum value.

Note in this figure that t_{CPA} for cycle 2 is triggered by the rising edge of CAS\ which also turns off the output buffers ending cycle 1.

The time for which the data is valid in this minimum cycle is the 5 ns time before the output buffers turn off (t_T) since t_{OFF} minimum is specified as zero. Some additional data valid time must be allowed in practice before the CAS\ goes high and turns the output buffers off since the 5 ns data valid time is very short to get a reliable reading.

The limiting expression for the minimum page cycle time in a real system becomes $t_{CPA} + t_T +$ reasonable data valid time. In our example this is 40 ns + additional data valid time. In a real system, the actual page cycle time in the system would probably need to be specified as at least 45 ns in the case considered.

Even given the limitations discussed here, a system where the data can be segmented by page can, theoretically, have a 40 ns fast page mode cycle time for random accesses within a given page of data. This compares favorably to the t_{CYC} of 110 ns for the full random access cycle for the -6 part.

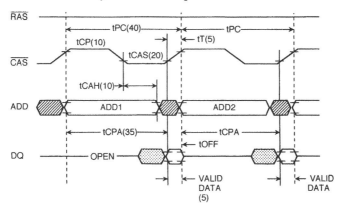

Figure 5.13   Example of fast page mode with maximum address set-up time (source: adapted from Micron Technology)

The speed for a -6 category ×1 DRAM in random access mode is 9.1 MHz and only 1.1MB/sec, whereas for the same part operating in fast page mode, the speed can be 25 MHz and the datarate is 3.1 MB/sec, a speed increase by a factor of 2.74 times.

If the DRAM also has a wide I/O, for example a ×16, then in random access mode the datarate is 18.2MB/sec and in fast page mode the datarate is 50MB/sec which is a 45-fold increase over the 1.1MB/sec datarate of a ×1 DRAM in random access mode. Clearly accessible datarate increases significantly from these early measures taken to increase the bandwidth of the DRAM.

Wide DRAM operation is illustrated for a 4M×4 fast page mode DRAM in the functional block diagram in Figure 5.14 where the various array blocks with their associated sense amplifier banks can be seen. Each of the four banks of sense amplifiers acts simultaneously as a page.

The maximum number of columns that can be addressed while in fast page mode is determined by the maximum RAS\ low pulse duration which is 100,000 ns in the example used. At a 40 ns page mode cycle time, all 2048 bits in the page can be accessed.

An additional advantage of page mode is that the power consumed during the page mode cycle is less than that consumed during a random access cycle. This is partially because the row address buffers and decoders do not need to be maintained in an active state during page mode access since only the column address buffers and decoders are involved but mainly due to the absense of sense and restore current during the page mode access.

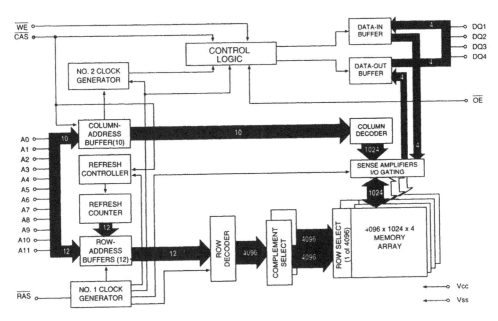

Figure 5.14   Functional block diagram of a 4M×4 fast page mode DRAM (source: Micron Technology)

5.4.3 Static Column Mode

Static column mode was the first of many attempts to make fast page mode even faster by reducing the number of required set-up, hold, or transition timings. It has not been widely used.

In static column operation the first access is normal with RAS\ clocking in the row address and CAS\ clocking in the column address. As with fast page mode, the RAS\ remains low for the succeeding cycles. In addition, the CAS\ is held low so that the column address buffer acts as a transparent flowthrough latch. When a new column address is applied, it flows on into the column decoder. The data then appears on the output after a time equal to the address access time (t_{AA}) as shown in the read cycle timing diagram in Figure 5.15.

As shown in the figure, while in static column mode, CAS\ can go high turning the output off, but when CAS\ transitions low again, the part remains in static column mode. When RAS\ returns high, the static column mode operation is terminated.

An early write cycle can be similarly achieved by toggling W\ as shown in Figure 5.16.

The minimum cycle time for static column is:

$$t_{SC} > t_{AA} + t_{AOH}$$

If we consider a -6 part similar in technology to the -6 part discussed in fast page mode, then for a minimum cycle with $t_{AA} = 30$ ns and $t_{AOH} = 5$ ns, then $t_{SC} = 35$ ns. The minimum data valid time is equal to t_{AOH} or, in this case, 5 ns which is the same as the fast page mode.

Recall that for minimum cycle time the fast page mode cycle time was 40 ns. Static column mode, therefore, saves one transition time over fast page mode.

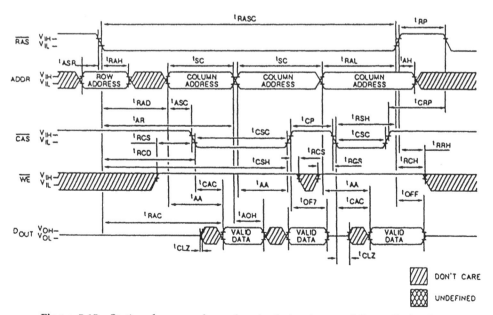

Figure 5.15 Static column mode read cycle timing (source: Micron Technology)

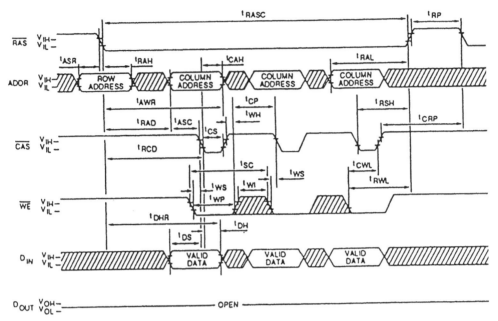

Figure 5.16   Static column mode early write cycle timing (source: Micron Technology)

The main drawback of static column mode is the lack of a defined clock for the output as was present in page mode. It is also more expensive for the chip manufacturers to implement an additional mode on the chip.

Most system designers have shown a preference for the added control of the clocked output and fast page mode became the most commonly used mode for fast DRAMs throughout the early 1990s.

5.4.4   Fast Page with EDO (Hyperpage)

Because of the limitations discussed above on the timings of the page mode cycle time, a mode with extended data out (EDO) has been defined. It is called Hyperpage Mode or EDO Page Mode.

Hyperpage mode is an unusual example of significant performance gain coupled with cost reduction. The various advantages are longer data valid time, faster access cycles, no additional silicon, no change in package or pinout, and the ability to achieve comparable speed to fast page mode with a lower cost technology.

The innovation with Hyperpage mode is to change the output path of the DRAM so that the rising edge of CAS\ no longer turns off the output buffers. This can be accomplished by putting a transparent flowthrough latch in the datapath after the amplifier circuits so the old data remains valid when CAS goes high. From the system perspective, Hyperpage mode allows column precharge to be hidden while latching the data out. As a result the minimum time CAS\ must be low is reduced and the rising edge can be earlier while data capture occurs in the system.

Hyperpage mode eliminates the t_{OFF} parameter since CAS\ no longer can take the output buffer to high impedance. Since the output buffer now remains in a low impedance state, a new timing parameter t_{COH} is defined which is the time from CAS\ low to data invalid. High impedance control of the output buffer can still be exercised by the OE\ or the WE\, or by RAS\ going high.

A simple comparison of fast page mode and Hyperpage mode read cycles is shown in Figure 5.17(a) and (b). These timing diagrams show that for both fast page mode and Hyperpage mode:

1. Row addresses are latched when RAS\ goes low.
2. Column addresses are latched when CAS\ goes low.
3. Column addresses are set up during CAS\ high time.

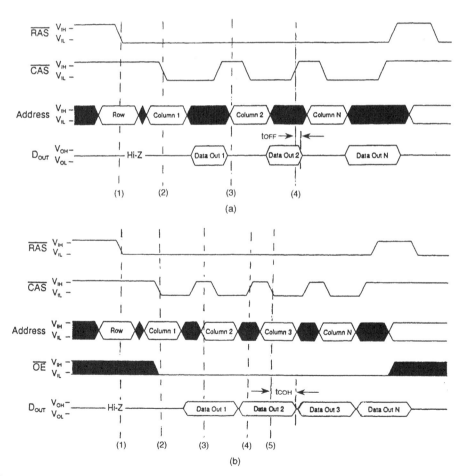

Figure 5.17 Comparison of (a) fast page mode; (b) Hyperpage mode read cycles (source: adapted from IBM)

The differences are:

4. In fast page mode (Figure 5.17(a)), data is disabled at a time t_{OFF} after CAS\ goes high, whereas in Hyperpage mode (Figure 5.17(b)) data stays valid when CAS\ goes high.
5. In Hyperpage mode, output goes to low impedance at a time t_{COH} after CAS\ goes low.

A read cycle for Hyperpage mode with the timing parameters indicated is shown in Figure 5.18. Note that the CAS\ goes high while the data is still valid and that the output buffers go to low impedance t_{COH} after CAS\ falls so that the internal column precharge is overlapped with the external data acquisition.

Since the data output drivers are not turned off when CAS\ goes high, a new address can be provided for the next access cycle before completing the current cycle. This ability to overlap cycles can give a significant performance advantage since it permits the overlap of limiting timing parameters such as t_{CPA}.

Being able to overlap timing cycles also allows shorter CAS\ pulse width and cycle times to be used. This can increase the speed even more while at the same time allowing the data to remain on the output longer than has been possible with fast page mode. With EDO there is a performance gain in overlapping the system's need for a valid data window with the DRAMs need for precharge.

An illustration of these performance gains can be seen by comparing the timing with maximum address set-up time for fast page mode, which is its fastest cycle, with that for Hyperpage mode.

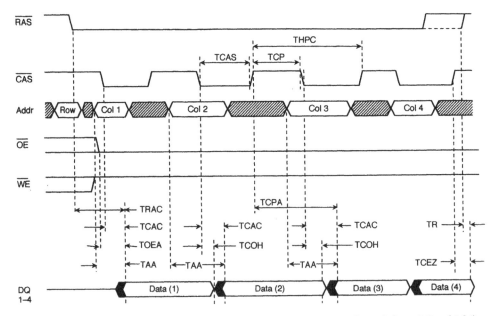

Figure 5.18 Read cycle timing for Hyperpage mode (source: adapted from Mitsubishi)

Recall that the defining equation for fast page mode was:

$$t_{PC(min)} = t_{CP} + t_{CAS} + 2t_T > t_{CPA} + t_T$$

where for fast page mode, t_{CPA} is the limiting parameter, and $t_{CPA} = t_{CP} + t_T + t_{CAS'}$ where $t_{CAS'}$ can be greater than $t_{CAS(min)}$. The capability of Hyperpage mode to overlap t_{CPA} permits $t_{CAS(min)}$ to be used.

Figure 5.19 illustrates the transition from fast page to Hyperpage (EDO) mode with the various advantages of Hyperpage shown step by step.

Figure 5.19(a) is essentially the same as Figure 5.13 and illustrates the minimum (fastest) cycle for fast page mode in a -6 technology. This cycle can be seen to be limited by $t_{CPA} + t_T$. The cycle time is 40 ns and the data is valid for a short 5 ns.

Figure 5.19(b) has the same timing parameters except the outputs are changed to an extended data out (EDO) type so that the rising edge of CAS\ no longer turns off the data. This extends the data to t_{COH} after the falling edge of CAS\. The data valid time is now a relaxed 25 ns and none of the other timing parameters have changed.

In Figure 5.19(c) the t_{CPA} is relaxed from 35 ns to 45 ns. Since the CAS\ precharge time (t_{CPA}) is optimized in the -6 technology for the fast page mode, it can reduce the cost of the technology to back off to a more relaxed t_{CPA} timing. This is possible since t_{CPA} is no longer a limiting parameter with EDO.

With the t_{CPA} relaxed to 45 ns, the data now becomes valid after the rising edge of CAS\ while CAS\ is high, sonething that was impossible in fast page mode. The cycle time remains 40 ns. A new cycle begins as always on the rising edge of CAS\ so that the cycles are overlapped by 5 ns. t_{CPA} is also overlapped with itself by 5 ns. The data valid time is still a reasonable 15 ns.

Finally in Figure 5.19(d), the t_{CAS} is shortened to 15 ns. Since the column address hold time, t_{CAH} is only 10 ns, and nothing else has to happen before CAS\ goes high, there is no problem with changing t_{CAS} to 15 ns. The result is that the cycle time is reduced to 35 ns and the data is valid to 10 ns. The t_{CPA} remains at its relaxed 45 ns value and now overlaps itself by 10 ns. Nothing else is changed.

The final result of these changes, which are made possible by EDO, is a shorter cycle time and a longer data valid time than in fast page mode while at the same time moving to a more relaxed technology.

This entire effect results from the fact that the rising edge of the CAS\ no longer turns off the output buffer so the data from one cycle can become valid after the CAS\ precharge for the next cycle has begun. Since the rising edge of CAS\ occurs before the data out is valid, this effectively permits the t_{CPA} timing to be overlapped from cycle to cycle. A more relaxed t_{CPA} timing becomes possible. The effect is both to shorten the total page cycle time, since the t_{CPA} is no longer a limiting parameter, and to increase the amount of time that the data can remain valid on each cycle, which improves the potential reliability of the system.

It is a major benefit of Hyperpage mode that the data remains valid longer than in fast page mode, in this case for 10/35 = 28 percent of the time for a -6 part for Hyperpage mode compared to 5/40 = 12.5 percent of the time for a similar technology fast page mode part. This advantage alone can mean a faster actual cycle time in the system using Hyperpage mode since additional data valid time will not be needed to ensure an accurate read.

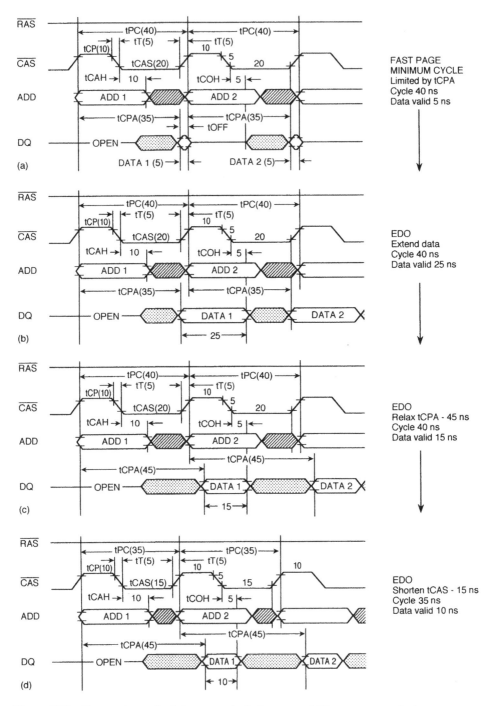

Figure 5.19   Comparison of minimum cycle fast page and Hyperpage mode (source: adapted from Micron Technology Memory Databook)

Figure 5.20 shows Hyperpage mode read cycle with maximum address set up time. This figure illustrates that after the addresses are latched in by the falling edge of CAS\, CAS\ can be pulled high and the next cycle begun before the data from the first access appears on the outputs. This is not possible in Fast Page Mode which must wait before pulling CAS\ high. This overlap permits a shorter cycle time.

The theoretical minimum cycle time for Hyperpage mode can be obtained by considering again the defining equations for fast page mode. They are:

$$(1) \; t_{PC} = t_{CP} + t_{CAS} + 2t_T$$
$$(2) \; t_{PC} = t_{CPA} + t_T$$
$$(3) \; t_{PC} = t_{AA} + 2t_T$$

The second equation is changed in the case of Hyperpage mode since the t_{CPA} is permitted to be overlapped by an amount $t_{CP} + \frac{t_T}{2}$. As a result, the second limiting equation for Hyperpage mode becomes:

$$t_{PC} = t_{CPA} - \left(t_{CP} + \frac{t_T}{2} \right) + t_T = t_{CPA} - t_{CP} - \frac{t_T}{2}$$

t_{CPA} is not normally a limiting parameter for Hyperpage mode as can be seen by putting in numbers from our previous examples.

The address access time t_{AA} is the limiting parameter in the case of Hyperpage mode (EDO). The minimum overlap permitted for t_{AA} in Hyperpage mode is about $t_T/2$ and equation (3) becomes:

$$t_{PC} = t_{AA} - t_T/2$$

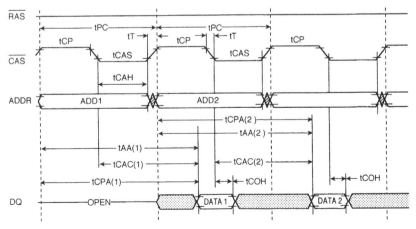

Figure 5.20 Hyperpage mode read cycle with maximum address set-up time (source: adapted from Micron Technology)

If we continue to use the minimum timing parameters for the -6 part, i.e. $t_{CP} = 10$ ns, $t_T = 5$ ns, $t_{CAS} = 15$ ns, with a relaxed $t_{CPA} = 45$ ns, and $t_{AA} = 35$ ns, the equations become:

$$(1)\ t_{PC} = t_{CP} + t_{CAS} + 2t_T = 35\ ns$$
$$(2)\ t_{PC} = t_{CPA} - t_{CP} - t_T/2 = 32.5\ ns$$
$$(3)\ t_{PC} = t_{AA} - t_T/2 = 32.5\ ns$$

If it would be possible to shorten t_{CAS}, t_T or t_{CP}, it would reduce the Hyperpage mode cycle time. Since $t_{CAH} = 10$ ns, it is possible to shorten t_{CAS} to 12.5 ns. For example if $t_{CAS} = 12.5$ ns, then $t_{PC}(1)$ would be 32.5 ns.

An example of Hyperpage mode read cycle with minimum address set-up time using these numbers is shown in Figure 5.21. It illustrates that the address access time (t_{AA}) is now the limiting access parameter.

The data valid time is 7 ns which, although shorter than in the previous example for maximum address set-up time, is still better than the comparable 5 ns minimum time with fast page mode.

It would be possible to further shorten the Hyperpage mode cycle time for a 3.3 V part by noting that the transition time t_T can be reduced to 2.5 ns, so $t_{PC(min)} = 12.5 + 10 + 5 = 27.5$ ns.

Beyond this minor adjustment, the minimum timing for hyperpage cycle has been reached. The limiting parameter now is effectively the address access time t_{AA}.

Table 5.2 compares the parameters for Hyperpage mode with those using fast page mode in the previous section using a part with the same -6 technology for comparison.

A further advantage of Hyperpage mode is that it potentially allows less costly components to be used to achieve a particular speed. In the example above, a -6

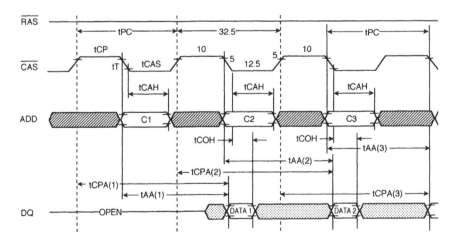

Figure 5.21 Read cycle for Hyperpage (EDO) with minimum address set-up time (source: adapted from Micron Technology)

Table 5.2 Comparison of minimum page cycle timing for fast page and Hyperpage mode for a 5 V -6 DRAM

Timing	Page (ns)	EDO (ns)
t_{CAS}	15	12.5
t_{CP}	10	10
t_T	5	5
t_{CPA}	35	45
. . .	. . .	. . .
t_{PC}	40	32.5
Data valid (max A)	5	10
Data valid (min A)	5	7

technology yields about a 32 ns Hyperpage part, so using a tighter technology would give an even faster Hyperpage equivalent. This assumes that other limiting parameters such as t_{AA} and t_{CAC} could be met.

Table 5.3 indicates such a comparison from the IBM 1995 DRAM Databook. The numbers are not exactly the same since parts differ from vendor to vendor due to differences in process and design, even in a similar geometry technology.

Hyperpage mode also has some minor design drawbacks. Design of the output stage is more difficult particularly on parts with wide outputs. The output has more ground noise and higher power dissipation. Also care must be taken if internal power supplies are used to drive the NMOS output pull-ups or these could be pumped down by a heavy output load.

Table 5.3 Comparison of fast page mode cycle and Hyperpage mode cycle for a given t_{RAC}

t_{RAC}	Fast page cycle	Hyperpage cycle
−60	40	25
−70	45	30
−80	50	35

Source: IBM 1995 DRAM databook.

5.4.5 Hyperpage Mode with Output Enable Control

Since it is frequently desirable to be able to put the outputs in the high impedance state (turn them off), Hyperpage mode timing can also have the output buffer controlled by the output enable (OE\).

Using OE\ control, it is possible both to turn off the output buffers and to restore the data from the previous cycle to the outputs as long as the falling edge of CAS\ has not occurred.

This is an additional advantage of Hyperpage mode, since with fast page mode it is not possible to restore the data once the outputs have gone into the high impedance state.

A simple illustration of Hyperpage mode timing with OE\ control is shown in Figure 5.22. The figure illustrates that the output buffer can be put in high impedance by a positive pulse of the output enable (OE\) while CAS\ is low which obeys the minimum OE\ hold time of t_{OEZ}. It also shows that the data from cycle 2 can be restored before another CAS\ falling edge occurs. However, after the CAS falling edge, the next data that will appear on the outputs is from cycle 3.

A positive going pulse of OE\ while CAS\ is high, after a time t_{OEZ} causes the outputs to become high impedance until the next CAS\ access.

The output is also shown going into a high impedance state t_{REZ} after RAS\ has gone high and t_{CEZ} after CAS\ has gone high.

5.4.6 *Hyperpage Mode with Write Enable Control*

The Write Enable pin can also be used in a Hyperpage mode read operation to control the outputs as shown in Figure 5.23. WE\ can put the output buffer into high impedance mode with a negative going pulse which obeys the hold time t_{WPE}. The output buffer turns off after time t_{WEZ} and is returned to a low impedance state by the falling edge of CAS\.

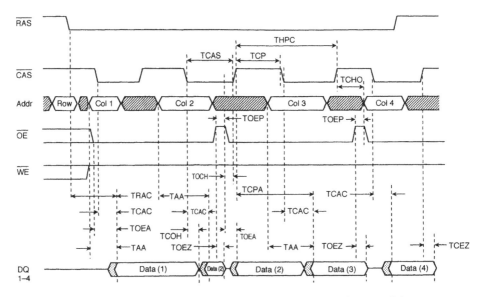

Figure 5.22 Hyperpage (EDO) mode read cycle with output enable control (source: Mitsubishi)

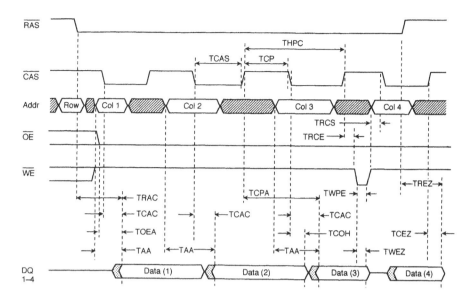

Figure 5.23 Hyperpage (EDO) mode read cycle using WE\ to control the outputs (source: Mitsubishi)

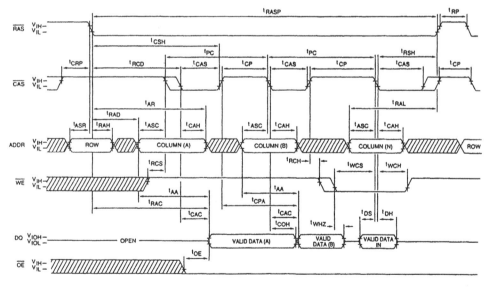

Figure 5.24 Hyperpage (EDO) mode RMW early write cycle (source: Micron Technology)

An example of the Hyperpage operation during a read–modify–write, early write cycle is shown in Figure 5.24. The figure first shows a read cycle (data A) with OE\ held active (low) and WE\ held high. The outputs go to low impedance a time t_{COH} after the CAS\ goes low and valid data (data B) appears after t_{CAC}. A write can then occur. The outputs with data B go to high impedance a time t_{WHZ} after WE\ goes low and a new address is set up and written in. After the write operation, the data outputs become high impedance until the next CAS\ access.

Summary of Hyperpage mode

Hyperpage mode gave DRAMs a way to bypass the constraint on timing from the CAS\ precharge access (t_{CPA}) timing parameter which had limited fast page mode.

The next challenge for the industry was to find a way to eliminate the constraint put on the Hyperpage mode timing by the address access time (t_{AA}). The following sections discuss various burst modes and pipeline burst mode options to eliminate this constraint.

4M VDRAM parts with EDO have also come out and will be discussed further in the next chapter.

Controllers for Hyperpage mode

It is important in the acceptance of a new DRAM access mode that controllers and chip sets be commercially available to support the mode. Fortunately Hyperpage mode has been supported by most of the major chip set manufacturers and appears on its way to becoming a generally implemented mode in the second generation 16M DRAMs and beyond. Because Hyperpage mode is so similar to fast page many existing controllers were able to adapt to the new mode by just shortening the $t_{CAS\backslash}$ timing.

5.4.7 Burst Mode with EDO

Burst mode with EDO enables data to be obtained without requiring a new address on every access, thereby eliminating the set-up time for an external address [8].

Burst mode works like a combination of Hyperpage mode and nibble mode. Only a single row and column address needs to be clocked in with the CAS\ and RAS\ and the rest of the column addresses in the burst are generated internally (like the older nibble mode which was a type of burst). Each data word in the burst must be clocked out with the CAS\ as with page mode. However, a new external column address is not needed after the first address.

A -6 DRAM is capable of 20 ns (50 MHz) burst cycle time compared to the 32 ns discussed earlier for Hyperpage mode without burst capability. A -5 part is capable of 15 ns (66 MHz) burst cycle time [8].

Burst mode uses an address increment counter to increment the column address which is latched into the column decoders on the falling edge of CAS\ during a burst cycle. The original address is maintained in the column address buffer. It is possible therefore to clock through an entire page of data with one set of external row and column addresses. The burst addresses wrap around upon reaching the end of the page.

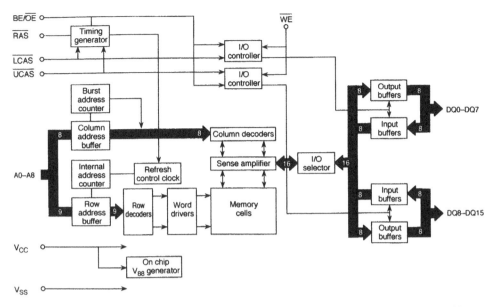

Figure 5.25   Functional block diagram of a 256K×16 burst EDO DRAM (source: Oki)

While the internally generated address needs some set up time, it is far shorter than that of an externally applied address.

A functional block diagram which indicates the location of the address increment counter of a 256K × 16 burst EDO DRAM is shown in Figure 5.25 [8].

The burst EDO DRAM shown here, when not in burst mode, is capable of functioning in both fast page and Hyperpage mode. The mode is controlled by the OE\ pin which in this case is also a burst enable (BE) pin. The part is in page mode when OE\ is high (the rising edge of CAS\ turns off the output buffer)and in Hyperpage mode when OE\ is low (the rising edge of CAS\ does not turn off the output buffer).

Timing diagrams for this part for fast page mode and Hyperpage (EDO) mode read cycle are shown in Figure 5.26(a) and (b).

A burst mode read cycle uses the EDO mode on the outputs as shown in Figure 5.27. Following the initial row and column addresses, the addresses are generated internally. Timing of the internally generated addresses is also shown for information.

The cycle begins with RAS\ and CAS\ latching in the row and column addresses. RAS\ is then held low throughout the burst. BE is also taken low at a time after the burst enable hold time is satified (t_{BEH}) and held low throughout the burst. After a time t_{CAC} valid data appears on the output. With BE low, the rising edge of CAS\ does not affect the data. The internal address counter increments as CAS\ goes high, setting up the next address, which is latched into the column decoder on the falling edge of CAS\. After a time t_{COH} the output buffer goes to low impedance. After a time t_{BAC} from the falling edge of CAS\ the next address becomes valid assuming that the CAS\ precharge access timing (t_{CBPA}) is met.

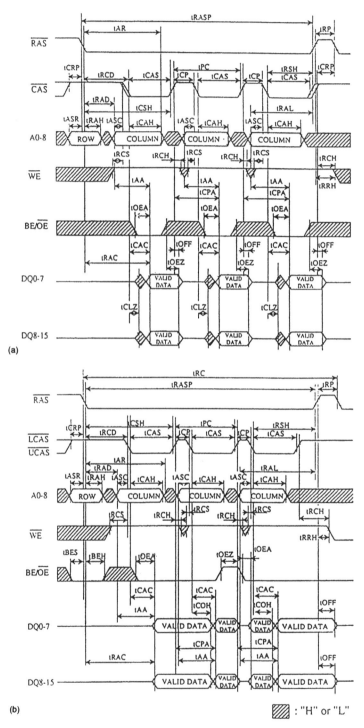

Figure 5.26   Read timing diagrams for a 256K×16 burst mode DRAM: (a) fast page; (b) Hyperpage (EDO) mode (source: Oki)

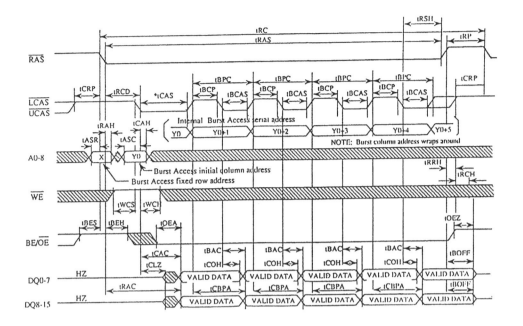

ₜ1CAS = 30ns (Min) at BURST READ TIMING

Figure 5.27 Burst EDO DRAM showing burst mode read cycle (source Oki)

This is no different from a normal Hyperpage mode cycle except that the addresses are internally generated. The addresses are provided so that the cycle has maximum address set-up time.

The minimum burst cycle timing is:

$$t_{BPC(min)} = t_{BCP} + t_{BCAS} + 2t_T$$

The constraints on the timing are t_{CBPA} t_{CAC} and t_{BAA} where we know that t_{CPA} can be overlapped for a Hyperpage mode cycle giving:

$$t_{BPC} = t_{BCPA} - (t_{CP} + t_T/2)$$

t_{AA} is the real constraint in Hyperpage (EDO) mode and was shown previously to be capable of being overlapped by one half a transition time giving:

$$t_{BPC} = t_{BAA} - t_T/2$$

Putting in numbers from the Oki datasheet for these parameters for a -6 ($t_{RAC} = 60$ ns) part (for comparison purposes with the earlier modes described) we have $t_{BCAS} = 8$ ns, $t_T = 1$ ns, and $t_{BCP} = 8$ ns, so $t_{BPC(min)} = 16 + 2 = 18$ ns. The datasheet specifies t_{BPC} at 20 ns.

Since the constraint on Hyperpage mode found previously for a minimum address cycle was $t_{BPC(min)} = t_{BAA} - t_T/2$ we can assume that t_{BAA} is no longer than 21 ns. If we look up t_{AA}, the address access for externally applied addresses, we find that it is

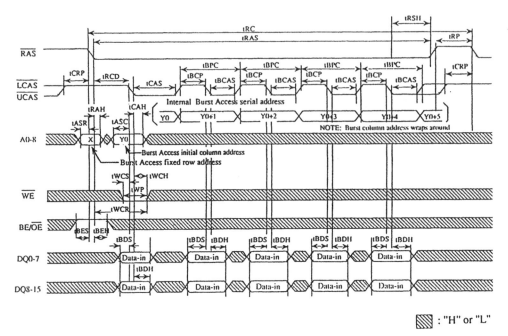

Figure 5.28   Burst mode write timing with EDO (source: Oki)

30 ns, which means that the address access for the internally generated addresses is significantly reduced from that required for an external address.

Figure 5.28 shows an EDO burst mode write timing diagram for the burst mode feature found on the same part.

Burst mode as described above is considered more useful for graphics DRAMs, which require pages of sequential information, than it is for main memory that requires random access be maintained.

5.4.8   Pipeline Burst EDO

A variation on the burst mode EDO is called pipeline burst EDO. Pipeline burst part shortens the bursts from a full page to four or eight bits and permits an external random address access for each burst without a timing penalty. The external address access time can be hidden in the burst cycle so that this limitation is removed from the minimum EDO timing.

The column address pipeline consists of an additional column address multiplexer included in the circuit between the column address buffer and the column decoder as shown in the functional block diagram in Figure 5.29.

The external column address is set up and latched from the column address buffer into the column address multiplexer on one CAS\ cycle and then into the column decoder on the next cycle. On subsequent CAS\ cycles the burst address counter increments the address in the column address multiplexer which is then latched into the column decoder by the falling edge of CAS\. On the last increment count (4 or 8),

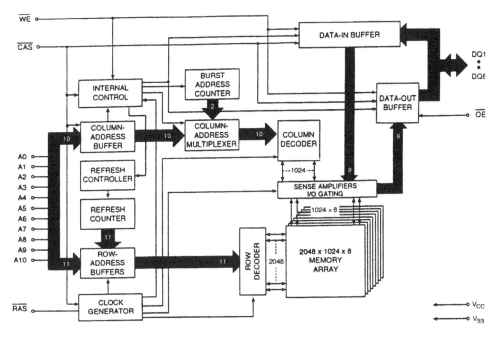

Figure 5.29   Functional block diagram of a pipeline burst EDO DRAM (source: Micron
Technology)

when the incremented address is set up on the column address multiplexer, a new
external address is also set up on the the column address buffer and on the next cycle
is latched into the column address multiplexer.

This pipeline permits a new random address to begin each burst without a gap.
The price that the pipeline burst DRAM pays is in an additional cycle of latency due
to the extra latch in the column address circuitry.

Both the linear and interleaved burst sequence of addressing can be provided by
the burst address counter. A WCBR cycle determines the burst sequence.

An example of a pipeline burst read timing diagram is shown in Figure 5.30.

For a burst read, a normal RAS\ and CAS\ cycle are followed by a single CAS
cycle to latch the column address from the column address multiplexer into the
column decoder. Subsequent CAS\ cycles on the falling edge of CAS\ both increment the burst counter to provide an address for the next cycle and latch in an
internally generated address for the current cycle.

As a result the page cycle timing (t_{BPC}) is equal to the address access time (t_{AA})
which, as we already know, is the minimum cycle constraint for Hyperpage mode
timing. As in the case of the normal burst EDO mode, the address access time is
significantly shorter with internally generated addresses than for those which are
externally generated.

Timings specified in this case [12] for the -6 specification ($t_{RAC} = 60$ ns) are:

$$t_{CP} = 5 \text{ ns, } t_{CAS} = 5 \text{ ns, } t_T = 1.5 \text{ ns}$$

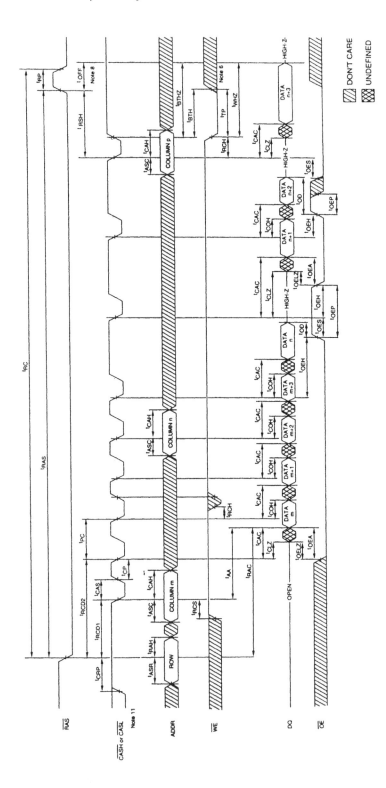

Figure 5.30 Pipeline burst read timing diagram (source: Micron Technology)

where

$$t_{PC(min)} = t_{CAS} + t_{CP} + 2t_T = 5 + 3 + 5 = 13 \text{ ns}$$

Actual t_{PC} is specified at 16.6 ns for the -6 part and 15 ns for the -5 part. The constraints are $t_{CAC} = 11.6$ ns, t_{AA} (external) = 30 ns and $t_{OEA} = 12$ ns.

It is clear that t_{AA} is still the limiting constraint for the external address since two 15 ns cycles are allocated to obtain a 30 ns t_{AA}. We can assume that t_{AA} for the internal address increment is 15 ns which is 66 MHz. In this case the optimum timing for the technology is specified for the -5 case.

The timing of the pipeline burst EDO DRAM and the burst EDO DRAM are similar. The advantage of the pipeline burst part is the ability to handle random addresses. The advantage of the burst EDO DRAM is no additional latency cycle and the ability to use the part as a standard fast page mode or Hyperpage mode DRAM.

The pinout and package of the pipeline burst EDO DRAM are not changed from the standard asynchronous DRAM and the chip size is increased by only the addition of the address counter which is minimal.

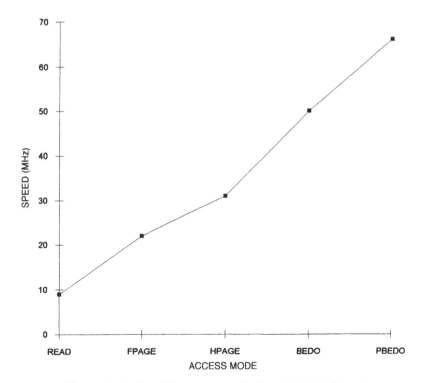

Figure 5.31 Speed by access mode for -6 DRAMs (MHz)

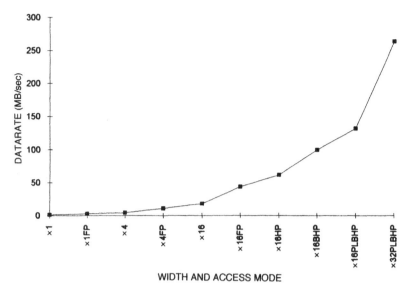

Figure 5.32   Datarate by width and access mode for -6 DRAMs (MB/sec)

The main cost of the pipeline burst is in the additional latency cycle at the beginning of the page and the cost of controller redesign needed to handle the new timings.

Theoretically pipeline burst can be used up to 66 MHz, a speed that was originally expected to be available only by going to the much more radical architectural changes in the synchronous DRAM.

A comparison of speed by access mode for the various DRAMs that have been discussed in this chapter is shown in Figure 5.31. An attempt has been made to use the -6 technology throughout for examples so that such a comparison could be made.

Since the system is more concerned about the rate at which data can be obtained from the DRAM, the datarate, Figure 5.32 shows a comparison of datarate by access mode and by width of I/O for the various access modes discussed.

Since databuses are commonly 64 bits or wider, the DRAMs described here can serve systems at datarates up to 500MB/sec.

It is expected that further modifications of the standard asynchronous DRAM for the purpose of gaining speed are unlikely and that at 100 MHz and above the industry will move to the Synchronous type of DRAMs which are discussed in the next chapter.

5.5   Technology Speed Trends

One factor that is helpful in improving the speed is that the technology tends to get faster as the DRAMs increase in density. This is due to many factors including shorter channel length transistors and thinner oxides.

Table 5.4 Typical datasheet parameters by density and technology

Technology (m)	0.4 64M/16M	0.6 16M/4M	0.8 4M/1M	1.2 1M
t_{CYC}	90	110	130	150
t_{RAC}	50	60	70	80
t_{RAS}	50	60	70	80
t_{CAS}	13	15	20	20
t_{PC} (page)	35	35	40	45
t_{CP} (page)	10	10	10	10
t_{CPA}	30	35	40	45
t_{AA}	25	30	35	40
t_T	3	3	3	5
t_{CAC}	8	10	15	15
t_{CP} (EDO)	10	10	10	10
t_{PC} (EDO)	23	25	33	35
t_{AA}	25	30	35	40
t_{CPA}	30	35	40	45
t_T	2	3	5	5

Source: various datasheets.

Table 5.4 shows the trend of various datasheet timings as the DRAMs increase in density both for fast page mode and for Hyperpage mode.

5.5.1 High Density 64M and 256M EDO DRAMs

The use of EDO (Hyperpage) mode has continued through the 64M and 256M DRAMs.

While mainstream computing systems are expected to be fully converted to SDRAMs by 2000, many industrial and mobile systems have continued using the asynchronous DRAMs. Factors such as requirements for low power DRAMs and for DRAMs that work with older processors will tend to keep the asynchronous DRAMs in production for a few years.

5.6 Other Factors in DRAM Speed

5.6.1 Access Time vs. Power Supply Voltage

Another factor which temporarily affects the speed trends of the DRAM is the migration to lower power supply voltages.

Typically if a DRAM designed for one power supply voltage is run at a lower power supply voltage, the speed will degrade about one speed selection. This was seen with the early 4M DRAMs which were originally in a 5 V technology.

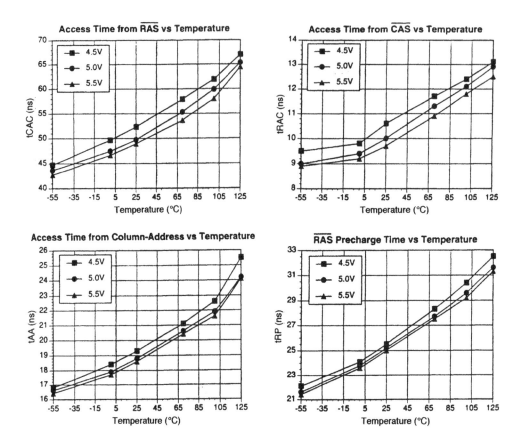

Figure 5.33   Timing parameters for a 5 V 4M DRAM vs. temperature (source: Micron Technology)

The 16M DRAM was designed in the 0.5m technology initially. In this technology it is necessary to have the internal array voltage at a lower value, around 3.3 V, to protect the circuitry.

Initially to meet 5 V demand for the 16M DRAM, a voltage converter was added to increase the voltage of the array to the 5 V required externally. When such a voltage converter is bypassed to change from 5 V to 3.3 V external voltage, the DRAM speed typically should not change significantly.

In many cases the 3.3 V version was actually slower by a speed sort anyway. This is because many of the early 16M DRAMs were actually designed in a slightly looser technology and used a 4 V array. A 5 V part in this case would slow down slightly in going to a 3.3 V external power supply. It is also easier to design a high drive buffer with 5 V than with 3.3 V. Similarly a 3.3 V 64M DRAM may have a lower internal voltage on the array.

5.6.2 Low Temperature Operation for Speed

Another method of speeding up DRAMs, which is not commonly used, would be to operate them in a lower temperature range than that for which they are normally specified.

For example parameters for a 4M DRAM are normally specified for a 0–70°C operating range. They are in each case specified with guardband at the worst case temperature. If the parts are maintained at a cooler temperature they will run faster as shown in the typical operating curves in Figure 5.33.

It appears that a 20 percent improvement in speed might be obtained by lowering the ambient temperature of the parts to 40°C lower than the maximum.

In a paper published in 1989, IBM [4] described the effect of a liquid nitrogen temperature environment on a 1Mb DRAM. The part had a 20 ns access time at room temperature (25°C) and in the liquid nitrogen environment it had a 12 ns access time.

5.6.3 Address Demultiplexed DRAMs

It is also possible to speed up a DRAM by demultiplexing the addresses. This was proposed by IBM in 1990. The drawback of this is that both the chip size and the package size increase, the chip size due to the increased number of address lines and the package because of the increased number of address pins. More recently Fujitsu has developed another fast address demultiplexed DRAM. Address demultiplexing has not been widely adopted for fast DRAMs.

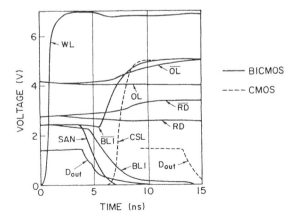

Figure 5.34 Simulated waveforms of CMOS and BiCMOS DRAM's at read operation (source: Toshiba [7] by permission of IEEE)

5.6.4 BiCMOS DRAMs for Speed

Another method of achieving a high speed DRAM is by the use of BiCMOS technology. Bipolar transistors have been used in peripheral circuitry requiring high drive capability to speed up the DRAM access time. Circuits such as address buffers, row decoders, and I/O sense amplifiers have been described in BiCMOS which can reduce the access time of the DRAM by as much as 30 percent [7].

For example, Hitachi [6] compared simulations of the performance of a BiCMOS driver with a CMOS driver for a 1Mb DRAM and found that the access times were 27.2 for the BiCMOS and 35.3 for the CMOS. The trade-off was that chip size of a 1Mb DRAM using BiCMOS drivers was about 10 percent larger than that of a CMOS DRAM in comparable technology.

More recently Toshiba [7] has described BiCMOS circuitry which significantly increases the speed of megabit DRAMS as shown in the simulated waveforms in Figure 5.34. This circuitry was implemented in a 1K DRAM BiCMOS test chip which had a 10 ns RAS access time compared to an 18 ns access time in a comparable 1K DRAM CMOS test chip. The area penalty is about 10 percent.

It remains to be shown that BiCMOS can be competitive with CMOS in deep submicron technology and lower than 3.3 V operating voltages.

5.7 Early Experiments in High Speed

Various DRAM test chips with high speed modes have also been presented at technology seminars in the past. Many of these are already described in Reference 3.

Bibliography

1. *IBM DRAM Databook*, 1995.
2. *Micron DRAM Databook*, 1994.
3. Prince, B., *Semiconductor Memories*, 2nd Edition, John Wiley and Sons, 1993.
4. Henkels, W.H. *et al.*, A 12 ns low temperature DRAM, *IEEE ISSCC Proceedings*, 1989.
5. *NEC Preliminary 16M DRAM Hyperpage Mode Data Sheet*.
6. Watanabe, T. *et al.*, Comparison of CMOS and BiCMOS 1Mb DRAM performance, *IEEE Journal of Solid State Circuits*, Vol. 24, No. 3, June 1989, 771.
7. Watanabe, S. *et al.*, BiCMOS circuit technology for high-speed DRAM's, *IEEE Journal of Solid State Circuits*, Vol. 28, No. 1, January 1993.
8   *Oki 4M Burst Mode DRAM Data Sheet*, 1994.
9. *Texas Instruments MOS Memory Databook*, 1993.
10. *Hitachi CMOS DRAM Databook*.
11. Mitsubishi papers on Extended Data Out Operation.
12. *Micron Pipeline Page Mode Databook*.
13. Prince, B., *Report on Single Port DRAMs for Graphics Subsystems*, Memory Strategies International, 1994.

6 New Architectures for Fast DRAMs

6.1 Overview

Thus far we have dealt with evolutionary changes made in the asynchronous DRAM to increase the effective datarate in the system. These changes have focused on widening the DRAM I/O structure and on improving the rate at which data from a single page can be obtained during a CAS\ cycle by minimizing the various internal timings of the DRAM in the critical path for data. The basic architecture of the DRAM has not been significantly altered.

Even with this speed and datarate improvement there has been an increasing mismatch in speed between the asynchronous DRAM and the new microprocessors. Due to design and architectural innovations, the microprocessors are expected to achieve speeds of up to 800 MHz by the end of the decade. This speed far surpasses that of the DRAMs even with the new burst modes described in the previous chapter which at most give speeds up to 66–80 MHz as shown in Figure 6.1.

This mismatch in performance has caused significant thought to be given to new architectures for DRAMs which can meet the required processor speed.

One promising direction has been to add a synchronous interface to the DRAM. Some devices which fall in this category include the Cache DRAM (CDRAM) from Mitsubishi, the JEDEC Standard Synchronous DRAM (SDRAM) and JEDEC Standard Synchronous Graphics RAM (SGRAM), and the synchronous DRAM from Rambus, Inc.

Another direction has been to interleave multiple independent banks on the DRAM chip. DRAMs using this technique include the JEDEC SDRAM and SGRAM, the DRAM from Rambus, Inc., and the MDRAM from Mosys.

A third direction has been to attempt to integrate the cache with the DRAM. Parts falling in this category include the CDRAM from Mitsubishi which has a separate SRAM and DRAM array on the chip and the Enhanced DRAM (EDRAM) from

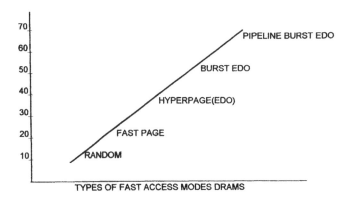

Figure 6.1   Asynchronous DRAM speed by access mode (MHz)

Ramtron which is asynchronous and uses distributed cache where static RAM cells share bit lines with the sense amplifiers.

A fourth category includes DRAMs which have a protocol logic and algorithm which can access a rapid burst of data. The DRAM from Rambus, Inc., and the Ramlink DRAM, a new IEEE standard which uses the IEEE SCI standard interface, fall into this category.

Some DRAMs use several of these techniques in addition to those described in the previous chapter in their attempt to increase their bandwidth in the system.

Figure 6.2 gives an indication of what is added to the DRAM array in these approaches in order of complexity.

The EDRAM adds the least silicon with only a distributed register. The one-bank synchronous DRAM, introduced by Samsung, adds a synchronous interface and mode register. The JEDEC synchronous DRAM then adds dual array banks and the DRAM from Rambus, Inc. adds logic protocol. The IEEE Ramlink Standard is similar to the part from Rambus but has a ring rather than a bus protocol.

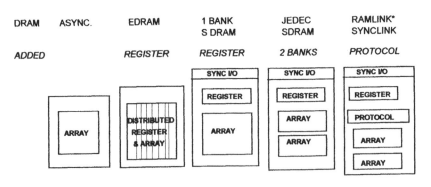

*The DRAM from Rambus Inc. is also of this type.

Figure 6.2   Evolution of fast DRAM architecture

Another direction, used mainly by video memories for CRT refresh, has been to try to access data in a fast serial stream. The frame buffer memory uses two serial ports to this end. The dual port DRAM or Video DRAM (VDRAM) adds, to the random port, a second port which is serial.

Synchronous interface on DRAMs will be considered in the following section together with a description of pipeline and prefetch techniques for speeding up a DRAM with synchronous interface. This will be followed by a description of multiple bank architectures, cache type DRAMs, and the protocol based DRAMs. Consideration of the graphics DRAMs will be taken up in the following chapter.

Examples of actual DRAMs which use each of these techiques are described later in the chapter.

6.2 Synchronous Interface on DRAMs

Historically DRAMs have been controlled asynchronously by the processor. This means that the processor puts addresses on the DRAM inputs and strobes them in using the RAS\ and CAS\ pins. The addresses are held for a required minimum length of time. During this time the DRAM accesses the addressed locations in memory and after a maximum delay (access time) either writes new data from the processor into its memory or provides data from the memory to its outputs for the processor to read.

During this time, the processor must wait for the DRAM to perform various internal functions such as precharging the lines, decoding the addresses, sensing the data, and routing the data out through the output buffers. This creates a "wait state" during which the high speed processor is waiting for the DRAM to respond thereby slowing down the entire system.

An alternative strategy is to make the memory circuit synchronous, that is, add input and output latches on the DRAM which can hold the data. Input latches can store the addresses, data, and control signals on the inputs of the DRAM, freeing the processor for other tasks. After a preset number of clock cycles the data can be available on the output latches of a DRAM with synchronous control for a read or written into its memory for a write.

Synchronous control means that the DRAM latches information from the processor in and out under the control of the system clock. The processor can be told how many clock cycles it takes for the DRAM to complete its task, so it can safely go off and do other tasks while the DRAM is processing its requests.

An advantage of synchronous DRAMs is that the system clock is the only timing edge that must be provided to the memory. This reduces or eliminates propagating multiple timing strobes around the printed circuit board.

For example, suppose a DRAM with a 60 ns delay from row addressing to data access (a -6 DRAM) is being used in a system which has a 10 ns (100 MHz) clock running as shown in Figure 6.3.

In the case of an asynchronous DRAM, as shown in Figure 6.3(a), the processor must apply the row address and hold it active while strobing it in with the RAS\ pin.

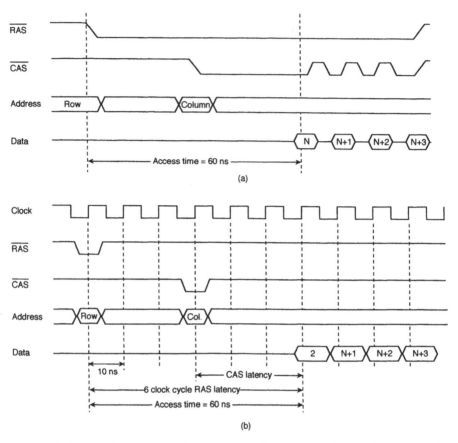

Figure 6.3 Comparison of DRAM with (a) asynchronous and (b) synchronous control

This is followed 30 ns later (t_{RCD}) by the column address which must be held valid and strobed in with the CAS\ pin. The processor must then wait for the data to appear on the outputs 30 ns later (t_{CAC}), stabilize, and be read.

For a DRAM with synchronous interface, however, the processor can lock the row and column addresses (and control signals) into the input latches, as shown in Figure 6.3(b) and do other tasks while waiting for the DRAM to perform the read operation under the control of the system clock. When the the outputs of the DRAM are clocked six cycles (60 ns) later, the desired data is in the output latches. The synchronous DRAM functions as a digital data storage device.

The underlying asynchronous DRAM has not changed. Specifically on the synchronous DRAM timing diagram in Figure 6.3(b), the row address and the CAS\ are latched in on the rising edge of the system clock activating a wordline. A column address is then clocked in three clock cycles (30 ns) later, after the minimum time required to activate the wordline (t_{RCD}). The data will then appear on the outputs after three more cycles (another 30 ns) to decode the column address and get the data from the sense amplifiers through the output buffers (t_{CAC}). Six clock cycles (60 ns), therefore, after the row address has been clocked in, the processor can expect to find the requested information on the output buffers of a synchronous DRAM.

Synchronous timing simplifies the inputs of the DRAM from the external perspective since all signals, addresses, and data can be latched in simultaneously rather than requiring the processor to monitor timings such as set-up and hold times. The output data is also simplified since the data will be in the output buffer latch on the appropriate cycle and the processor needs only to clock it out. The complex timing diagrams of the underlying asynchronous DRAM operation are not required to use the synchronous DRAM in a digital system. A knowledge of the command truth table and clock latencies are all that is required.

6.3 High Speed Modes on Synchronous DRAMs

Similar techniques to those described in the last chapter for the asynchronous DRAM can be used to hide the different components of the internal timing delay on a synchronous DRAM. For example, the active command to read/write command, t_{RCD}, and precharge time, t_{RP} can be hidden after the first access by using a page mode or a burst mode in which a series of data bits from the page currently on the sense amplifiers can be clocked out rapidly following the access of the first bit.

For a page mode a new column address is needed for each access. For a burst mode, following an initial address input, subsequent addresses are internally generated in rapid succession without inputting new address information to the DRAM.

An example of a burst mode on an ordinary DRAM was shown in Figure 6.3(a) where four bits of data from locations N, $N + 1$, $N + 2$ and $N + 3$ are clocked by the falling CAS\ onto the data lines following addressing of location N.

Burst mode functionality can be extended beyond the four bits up to a full page of data. Synchronously controlled burst mode is foreseen as being useful for high speed main memory functions as well as video support requiring 100–200 MHz datarates.

Burst mode can also be generated with linear or interleaved address sequences. An example of some linear and interleaved address sequences from an eight-bit burst is shown in Table 6.1. This is useful since different processors use the different burst mode sequences. For example, the Intel processors use the interleaved mode and the Motorola processors use the sequential mode.

Burst can be combined with a "wrap" feature as shown in Table 6.1which allows the string of bits stored both before and after the initial bit location in the DRAM to be accessed in rapid succession after the initial access. This feature is expected to be useful for high speed cache fill applications since the most likely bits to be wanted next are those physically close to the given bit.

Table 6.1 Linear and interleaved address sequences for an 8-bit burst

Initial address	Sequential	Interleaved
000	01234567	01234567
011	34567012	32107654
110	67012345	67452301

These features can be made programmable, as on the JEDEC Standard Synchronous DRAM, by use of a special configuration cycle with which the user can program both the type of wrap desired as well as the number of bytes available on each wrap.

6.4   Pipelining on Synchronous DRAMs

A synchronous DRAM architecture also makes it possible to speed up the average access time of the DRAM by pipelining the addresses. In this case it is possible to use the input latch to store the next address which the processor will want while the DRAM is operating on the previous address. Normally the addresses to be accessed are known several cycles in advance by the processor. Therefore, the processor can send the second address to the input address latch of the DRAM to be available as soon as the first address has moved on to the next stage of processing in the DRAM. This eliminates the need for the processor to wait a full access cycle before starting the next access to the DRAM.

An example of pipelining is shown in the schematic diagram in Figure 6.4 The pipeline is the column address path and is the more critical path for this example since in normal operation the row is held active and a page mode is used. [9]

The column address-to-output path is a three-stage pipeline. The address buffer is the first latch. A column switch is the second latch and the output buffer is the third latch. The latency inherent in the column access time is therefore divided up between these three stages.

The timing operation for a pipelined read is shown in Figure 6.5. The column address (A1) is clocked into the address buffer on one cycle and is decoded. On the second cycle the column switch transfers the corresponding data (D1) from the sense amplifier to the read bus, and column address A2 is clocked into the address buffer. On clock three, the data (D1) is clocked into the output buffer, D2 is transferred to the read bus, and A3 is clocked into the column address buffer. When D1 appears on the output, D2 and D3 are in the pipeline behind it.

While it takes three cycles for the data from A1 to appear on the output buffer, it only takes one more cycle for the data from A2 to appear or only one third as long. This means that pipelining can reduce the CAS cycle time for a page if addresses are known several cycles in advance. This is particularly useful for a burst mode where the subsequent addresses are known in advance because they are generated internally.

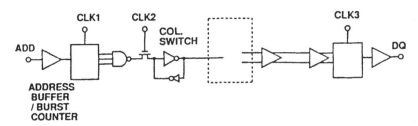

Figure 6.4   Example of a three-stage column address pipeline: (source: T. Yasuhiro (1994) [9] permission of IEEE)

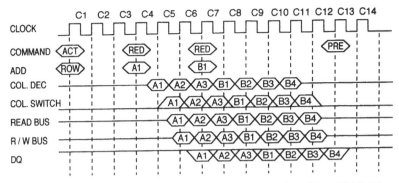

Figure 6.5 Timing operation for a pipelined read (source: T. Yasuhiro (1994) [9] permission of IEEE)

Another advantage is that a pipelined architecture does not increase the chip size significantly since it uses the DRAM's conventional cell array and the DRAM's normal operation fits well into the pipeline structure. There are no additional address or read buses required and also no additional read/write bus wiring.

It also permits interruptions of a burst on any address since both the external address and the internally generated address are transmitted through the pipeline in the same manner. This permits a random column access with a different external address on every clock cycle.

One of the drawbacks of pipelining is that there is addressing on every cycle so that all stages of the DRAM are active at the same time. This can cause the power consumed during operation to increase over that of a DRAM where only one stage is active at a time. There is also a potential for increased disturbance problems due to the additional internal noise during the sensing operation.

In addition, with pipelining the part must be run at a reduced operating frequency due to the need to precharge long I/O lines during read. Also for a write the column select line has to remain at a high potential long enough to write data from the I/O lines to the bitlines but still stay low long enough to precharge the I/O lines without contention in order to read quickly.

6.5   Prefetch Architectures in Synchronous DRAMs

Another technique for increasing the speed of a synchronous DRAM is called prefetch. In this case more than one data word is fetched from the memory on each address cycle and transferred to a data selector on the output buffer. Multiple words of data can then be sequentially clocked out for each address access.

The main advantage of this approach is that, for any given technology, data access can be at multiples of the clock rate of the internal DRAM.

Another advantage of this architecture, since the speed can be obtained without clocking the internal architecture at as high a rate as the outputs, is that both the power dissipation and noise can be reduced for long sequential bursts.

There are also some drawbacks to a prefetch type of architecture. An output register must be added to the chip to hold the multiple words that are prefetched which adds to the chip size. If more than two address bits (two data words) are prefetched, it adds considerably to the chip size but ensures a fast unbroken data stream that can help hide precharge and RAS and CAS accesses.

An eight-bit prefetch scheme, for example, can achieve a very high frequency of operation for long bursts but it adds a considerable amount of chip area. In addition the power can increase if random addressing is required due to data thrashing.

A two-bit prefetch scheme adds an acceptable amount to the chip size for narrow I/O width memories. For wide word I/O memories, such as ×32 I/O widths, even a two word prefetch scheme may have an unacceptable die size penalty. With a two-bit prefetch, however, there are limitations on the timing. New column addresses can only occur on alternate cycles since there are always two address bits generated for every access.

Examples are shown in the block diagrams in Figure 6.6 of pipelined and prefetch output structures. Figure 6.6(a) shows a two-bit prefetch output structure, and Figure 6.6(b) shows a pipelined output buffer.

6.6   Combinations of Pipelining and Prefetch

Combinations of pipelining and prefetch are also used. For example, a 16Mb synchronous DRAM with a 125 MHz output clock [13] has been developed. This 2M×8 organized part combined two-bit address prefetch with a pipelined address lookahead. The I/O transfer scheme and column select line structure are shown in Figure 6.7.

There are four I/O's and each column select line picks two of them at a time. Since consecutive column selects pick different I/O lines an overlap is permitted which gives additional operating margin to complete writing onto the bit lines at a high datarate. Also the unselected pair of I/O lines can be precharged while the other pair is being accessed in a read cycle. The externally addressed bit is accessed while the internal column address counter is updated for the next address.

Therefore with a two-bit prefetch the column select access begins one clock cycle ahead of the external burst clock. If the burst is interrupted with a row precharge, the

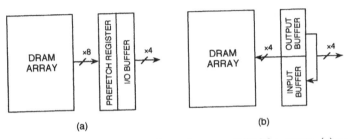

(a)   (b)

Figure 6.6   Block diagrams of two types of synchronous DRAM outputs: (a) prefetch; (b) pipelined

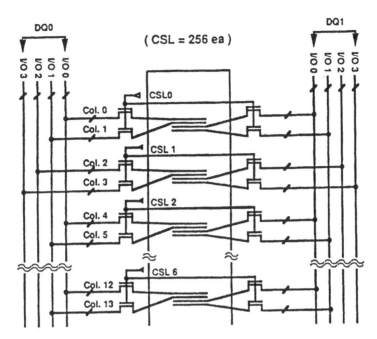

Figure 6.7   16M SDRAM with two-bit prefetch and pipelined address lookahead (I/O and column select line scheme) (source: Choi (1994) [13] by permission of IEEE)

last two bits already on the I/O lines can be read out to help hide some of the precharge time.

6.7 Multiple Internal Banks

Multiple independent internal banks can also be used on a DRAM. These banks can permit faster random access by allowing one bank to precharge or be refreshed while another bank is being accessed.

An early example of hiding the precharge time by interleaving two banks on an asynchronous DRAM was the "nibbled-page" architecture from Toshiba [10]. In this device the data from eight bit sections of different columns was interleaved on chip to give byte level random access at a 100 Mbit/sec rate. Nibbled page mode is described further in Reference 4.

A block diagram of a JEDEC Synchronous DRAM showing a two bank architecture is shown in Figure 6.8.

Multiple internal banks can help small fast systems with the memory granularity problem. Additional speed can be achieved by "ping-ponging" between the banks on one chip rather than by interleaving multiple banks in the system which can increase the cost by adding unneeded memory. The penalty which is paid over a one bank architecture is in power since both banks must be kept active.

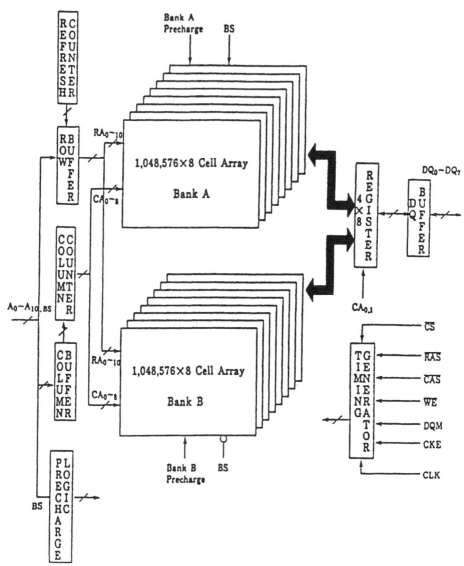

Figure 6.8   Block diagram of an SDRAM showing the multiple bank structure (source: *Toshiba* [12])

Two banks can help hide some of the row address precharge time when random access is required during burst mode sequences. Burst mode requires that all the bits to be accessed are physically located in the same row of cells as the initial access. If data from different rows is necessary as in a random access, it is possible to hide some of the row address precharge time if the DRAM has multiple independent banks and the two rows lie in different banks.

There are two possible modes of operation to hide some precharge time by alternating banks. One bank can be precharged while the other bank is active or it is also possible to keep both banks active and "ping-pong" between them.

This method of interleaving banks increases the DRAM speed in the system by cutting the effective access time. The requirement that sequential addresses lie in different banks can, however, require the system to use more memory than is actually needed in the particular system for data storage and it makes the system control more complex. In the case where both banks remain active the power dissipation is higher than when one bank is idle.

The other advantage to multiple bank DRAM architectures is that the active rows act as a cache. Once an access to a row in a given bank is made, that full row is then held active and can be accessed again simply by supplying a new column address. Having rows active in two different banks doubles the effective cache size.

An example of two-bank operation for a 16M SDRAM is shown in the read cycle timing diagram in Figure 6.9 for a four-bit burst with wrap mode[12].

At clock cycle 0, a row address is presented for bank A along with a 'bank A Address command. BS is the pin which selects the bank. At cycle 3, the column address is presented for bank A in a 'bank A Read' command. Even before the data is available from bank A, a row address for bank B is presented in clock

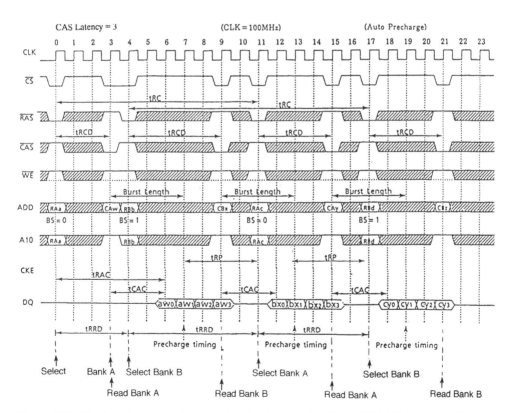

Figure 6.9 Read cycle timing showing two-bank operation in an SDRAM (source: adapted from Toshiba [12])

cycle 4 with an address command. The data from bank A begins to appear on the outputs in clock cycle 6. The column address for bank B is presented in cycle 9 with a 'read bank B' command, and the data from bank B appears in cycle 12.

An autoprecharge command for Bank A was issued in cycle 3 by holding A10 high. This prepared it to be ready to accept a new row address at a later time. (There is further explanation of the autoprecharge command in Section 6.12.2.)

Notice there are two cycles between data from the A row, and data from the B row at clock 10 and 11. The break in data is caused by the row-to-row delay required being longer than the burst length. If an eight-bit burst had been used the gap in the data could be eliminated. It is possible to allow continuous data from either bank throughout the entire *row* if a long enough burst is used.

6.8   Overview of Types of Synchronous DRAMS

Early synchronous DRAMs began appearing on the market commercially in 1994. These included: the 16M JEDEC standard synchronous DRAMs (SDRAM) from several suppliers, most of which had two-bank architecture; a similar 16M part from Samsung which had level RAS input rather than the pulsed RAS used by the JEDEC part and had only one bank; a 4M and 16M DRAM with SRAM cache on the chip (CDRAM) from Mitsubishi; and a 4M and 18M synchronous DRAM with a special bus protocol from Rambus, Inc. There were also a number of synchronous DRAM types which were intended for the graphics subsystem market and will be covered in the next chapter.

The following section looks more closely at these different early commercial products.

6.9   The Early 16M JEDEC SDRAMs

6.9.1   Features

The features that appeared on the early JEDEC type SDRAMs are shown in Table 6.2.

Some additional non-standard features were also included on many JEDEC Standard parts. For example, some vendors also offered bursts of one-bit or full page burst. Vendors with pipelined internal architecture offered the option to assert random column address in every cycle. These features were not included on the JEDEC standard since the intent of a standard is to provide a minimum set of features that must be on all parts claiming to meet the standard.

If one feature of the minimum standard is missing on a part, however, it can not claim to meet the standard. For example, one of the early SDRAMs used a level RAS instead of a pulsed RAS as called for in the standard and thereby did not meet the standard. Some vendors also offered various low swing interfaces on the JEDEC SDRAM such as GTL and CTT, and at least one vendor had autosensing

Table 6.2   Feature summary of early 16M JEDEC type SDRAM

- Fully synchronous with all signals referenced to a positive clock edge.
- Organization: dual internal banks, ×4, ×8, ×9, ×16, ×18
- Package: 400 mil, 0.8 mm, 50-pin TSOPII (JEDEC Standard)
- Power supply: 3.3 V, 10%
- Interface: LVTTL
- Programmable burst length: 2, 4, or 8
- Programmable output sequence: serial or interleaved
- Control pins: Chip Select and Clock Enable
- DQ Mask Function
- Byte control using LDQM/UDQM on ×16, ×18 option
- Programmable read latency from column address
- CBR auto refresh
- Self refresh during power down
- Pulsed RAS
- Random Column Address on alternate cycles during burst
- Random

between LVTTL and GTL interface on their first part. Low swing interfaces will be discussed in a later section. Most of these early specifications listed speeds from 50–100 MHz.

6.9.2   New Pin Function Descriptions

The pinouts of the JEDEC 16M SDRAM is shown in Figure 6.10. Figure 6.10(a) shows the 44-lead pinouts of the ×4, ×8 and ×9 organized 16M SDRAM and Figure 6.12b shows the 50 lead pinout of the 1M×16 SDRAM.

The pins that are shown on these pinouts are defined in Figure 6.10c. Various new pins and pin functions are defined for the JEDEC SDRAM. These are:

CS\ – Chip Select: The CS\ pin enables all command inputs including RAS\, CAS\, WE\, and Address. When CS\ is high command signals are ignored but internal operation such as a burst cycle will not be suspended. In a small system CS\ can be tied to ground as with a normal chip select.

BS – Bank Select: The SDRAM can have multiple banks. Most 16M SDRAMs have two banks. Bank selection occurs at the "Bank Active" command followed by a read, write, or precharge command.'Bank Select' was later changed to 'Bank Address' by JEDEC.

DQM – Data-in Mask and Output Enable: The DQM (L and U) are active high.

CKE – Clock Enable: The CKE pin is used both as the Clock Enable, to suspend the clock, and to put the SDRAM into a power down state.

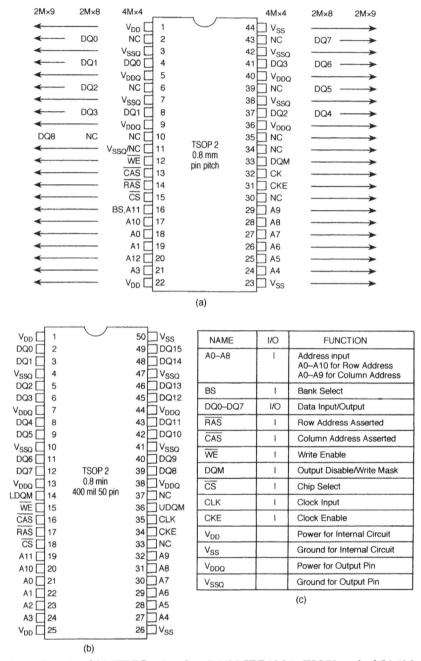

Figure 6.10 Pinouts of (a) JEDEC ×4, ×8, ×9 16M SDRAM in TSOP2 and of (b) 1M×16 in TSOP2; (c) pin definitions

6.10 Architecture of JEDEC SDRAMs

6.10.1 Synchronous and Registered Inputs and Outputs

Synchronous DRAMs have latches on all input, output, and control pins. The cost of the input and output latches is in adding to the initial latency of the DRAM. After this initial first access penalty is paid, however, the time to access subsequent data in the output latch can be a matter of a few nanoseconds.

A dual latch is called a register. Latches permit the data which is clocked into it to flow on beyond the latch. A register clocks in and holds the data until the next clock [4]. The addition of a latch to a latched input creates a register.

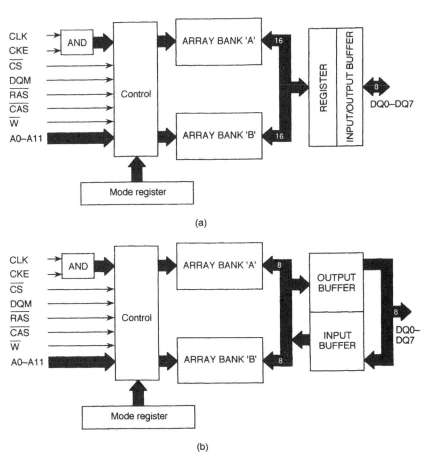

Figure 6.11 Block diagrams of two types of early JEDEC SDRAM architectures: (a) prefetch (2:1); (b) pipelined

6.10.2 Multiple Internal Banks

The SDRAMs are defined to permit multiple internal banks. This permits the parts to hide precharge and to interleave between two banks. Multiple rows on this part can be simultaneously open and accesses can be interleaved on the chip between the two banks.

6.10.3 Output Structure

There are two types of output structures available on the SDRAM – those with multiple word output registers ("prefetch type") and those without these registers ("pipeline type"). These were described in Sections 6.4 and 6.5.

Functional block diagrams of the two types of internal architectures are shown in Figure 6.11 where the two-bank internal structure as well as the two types of output structures can be seen.

The parts with multiple word output registers can hold either two, four, or eight words. A 2:1 prefetch register is shown. These output buffers can interleave the data out and hence run the output at multiples of the speed of the array. The advantage is that for any given array technology, the output can be run at a higher speed than the array. The drawback of the two-word architecture is that back-to-back CAS\ addressing during a burst is not possible. This has given rise to a "2–N rule" (see Section 6.12.7) in the JEDEC Standard. The drawback of the four-word and eight-word buffer architectures is the extra silicon consumed.

The parts without multiple-word output registers are generally pipelined internally to improve speed. The output in this case runs at the speed of the array, rather than in multiples of it. The pipelined architecture has the advantage of allowing back-to-back CAS\ accesses (which are not part of the JEDEC standard). Its drawback is the need to use faster, more expensive technology to match the output speed attainable by the parts with prefetch architecture.

6.11 Operational Features

6.11.1 Mode Register

The JEDEC SDRAM is a state machine controlled by the system clock. All internal strobes required for DRAM operation are referenced to this external clock. The SDRAM is controlled by programming the DRAM modes of operation into a mode register which is programmed after power-on and before normal operation. In addition, the mode register may be changed during operation.

Data contained in the mode register includes burst length, burst sequence type, CAS\ latency, and the operational mode (whether normal operation or test mode).

The "mode register set cycle" is initiated by a clock rising edge which occurs while holding CS\, RAS\, CAS\, and W\ low and having present on the address lines the

a)

A9 A8	A7	A6 A5 A4	A3	A2 A1 A0
OPCODE	O	L M O D E	BT	BL

b)

A9	A8	WRITE MODE
0	0	BURST READ & BURST WRITE
0	1	RESERVED
1	0	BURST READ & SINGLE WRITE
1	1	RESERVED

c)

A6	A5	A4	CAS LATENCY
0	0	0	RESERVED
0	0	1	1
0	1	0	2
0	1	1	3
1	X	X	RESERVED

d)

A3	BURST TYPE
0	SEQUENTIAL
1	INTERLEAVE

(e)

A2	A1	A0	BURST LENGTH	
			BT=0	BT=1
0	0	0	1	1
0	0	1	2	2
0	1	0	4	4
0	1	1	8	8
1	0	0	R	R
1	0	1	R	R
1	1	0	R	R
1	1	1	F.P.*	R

*Full Page

Figure 6.12   Block diagram of 16M SDRAM mode register

valid mode information to be written into the register. At least two "no operation" cycles must follow a mode register command.

The five part architecture of the 16M (18M) SDRAM mode register is shown in the block diagram in Figure 6.12(a).

The OPCODE field is A8–A9 and is used to set the write mode as shown in Figure 6.12(b). A7 is the mode register select with the register shown selected for A7 = 0 which is normal operation. If A7 = 1 then the part is in test mode.

The LMODE field, which sets CAS latency, is A4–A6 as shown in Figure 6.12(c). CAS latency may be set for 1, 2, or 3.

The burst type is set with A3 as either "0" for sequential mode or "1" for interleave mode as shown in Figure 6.12(d).

The burst length is set with A0, A1, and A2 as shown in Figure 6.12(e). Burst lengths permitted by the standard are 1, 2, 4, and 8 and full page. A burst length of 1 is similar to a standard asynchronous DRAM and permits testing with the same test equipment.

These mode register functions are described in more detail in the next section.

6.11.2 Burst Mode Access

The synchronous DRAMs have burst mode accesses which are compatible with the Intel 486 processor and later generation processors. These very fast accesses of bursts of data follow an initial access at the normal speed of the memory. The addresses of the subsequent bits of data in the burst are generated automatically by the RAM.

These burst mode accesses, like page mode, take advantage of the fact that the internal bus of the RAM is wider than the external bus. This permits all of the data from a series of burst mode addresses to be rapidly fetched from the RAM databank to its outputs after the entry of the first address. This data can then be fed out of the RAM at the speed of the output buffer.

The length of the burst is specified in the first three bits of the mode register and can be either 1, 2, 4, 8 or full page.

The burst type can be either sequential mode, like the Motorola processors, or interleaved mode, like the Intel processors.

The sequential mode is an incremental decoding scheme which wraps to the least significant bit as shown:

Sequential burst mode:	Burst length		Address		
	1	2	3	4
	2	1	0	HI–Z	HI–Z
	4	1	2	3	0

The interleaved mode is a scrambled decoding scheme interleaving increments of even and odd as shown:

Interleaved burst mode:	Burst length		Address		
	2	0	1		
	4	1	0	3	2

The burst type is also programmed in the mode register as shown in Figure 6.12.

6.11.3 CAS Latency

A burst can occur after a sense amplifier bank on the SDRAM has been accessed in a Bank Activate and Row Address Strobe operation which is similar to the RAS\ access of the asynchronous DRAM. This is referred to as "opening the bank" on an SDRAM.

To begin a burst read cycle on the SDRAM, the addresses are set up and RAS\ and CS\ are held low on the rising edge of the CLK signal. This activates the designated bank of sense amplifiers as shown in the READ timing diagram in Figure 6.13.

After a number of clock cycles equal to the minimum RAS\ to CAS\ delay (t_{RCD}), the CAS\ and CS\ are held low and strobed in on the rising edge of the CLK signal. After clock cycles equal to the column access time (t_{CAC}) the first bit of data is on the output and can be retrieved on the rising edge of the CLK pin. The access depends on

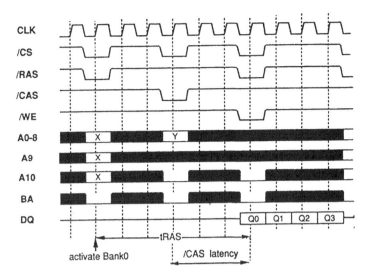

Figure 6.13 Read cycle timing diagram for SDRAM showing RAS\ and CAS\ latency (source: Mitsubishi [20])

the CAS latency programmed, but must obey the rule that CAS latency (CL) times t_{CLK} must be longer than t_{CAC}, that is CL $\times$ $t_{CLK} \geq t_{CAC}$. On subsequent clock cycles the rest of the burst of data can be clocked out. The data burst in the case shown is four bits long.

The CAS\ latency of the first address in a burst can be specified as 1, 2, or 3 clock cycles depending on the system clock frequency and the intrinsic maximum CAS\ access time (t_{CAC}) of the DRAM. CAS\ latency is programmed in the mode register.

6.11.4 Chip Select Latency

The Chip Select Latency (CS\) for the JEDEC SDRAM is zero. For higher speeds this could be a noisier option than a CS\ latency of one which would allow the addresses and RAS to be set up and clocked in before the chip is selected.

6.12 Operational Functions of the SDRAM Truth Table

6.12.1 SDRAM Truth Table

The various SDRAM operational functions are similar to those of the standard DRAM. The input and output states required to execute these functions are shown in the simplified functional truth table in Figure 6.14. States not shown in this

SDRAM FUNCTIONAL TRUTH TABLE

COMMAND	CKE	DQM	A11	A10	A9	A8–A0	CS\	RAS\	CAS\	WE\
MODE REGISTER SET	H	X	V	V	V	V	L	L	L	L
AUTO-REFRESH	H	X	X	X	X	X	L	L	L	H
SELF-REFRESH ENTRY	L	X	X	X	X	X	L	L	L	H
SELF-REFRESH EXIT	H	X	X	X	X	X	L	H	H	H
PRECHARGE SINGLE BANK	H	X	V	L	X	X	L	L	H	L
PRECHARGE ALL BANKS	H	X	X	H	X	X	L	L	H	L
BANK ACTIVE (RAS)	H	X	V	V	V	V	L	L	H	H
WRITE	H	X	V	L	X	V	L	H	L	L
WRITE WITH AUTO-PRECHARGE	H	X	V	H	X	V	L	H	L	L
READ	H	X*	V	L	X	V	L	H	L	H
READ WITH AUTO-PRECHARGE	H	X*	V	H	X	V	L	H	L	H
RESERVED	H	X	V	V	V	V	L	H	H	L
NO OPERATION	H	X	X	X	X	X	L	H	H	H
DEVICE DESELECT	H	X	X	X	X	X	H	X	X	X
CLOCK SUSPEND	L	X	X	X	X	X	X	X	X	X
POWER DOWN	L	X	X	X	X	X	X	X	X	X
DATA WRITE/OUTPUT ENABLE	H	L	X	X	X	X	X	X	X	X
DATA MASK/OUTPUT DISABLE	H	H	X	X	X	X	X	X	X	X
MODE REGISTER READ	H	X	V	V	V	V	L	L	L	L
TEST MODE	H	X	V	V	V	V	L	L	L	L

NOTE: V=VALID, L=LOGIC LOW, H=LOGIC HIGH, X=EITHER L OR H
*For CAS latency >2

Figure 6.14   Functional truth table for SDRAMs

diagram are illegal states. A "no operation" (NOP) state is one for which no operation has been defined.

6.12.2 Auto-Precharge

It can be specified that the bank currently being accessed shall precharge itself as soon as the burst is complete. This is done on the 16M SDRAM using address bit A10 during the column address portion of any cycle. If A10 is low the bank is not precharged but is left active at the end of the burst. If A10 is high, a bank is precharged automatically at the end of the burst. The specific bank is specified by the state of All. Another command can not be issued until the minimum timing for auto precharge has elapsed. This can be the same as or less than the minimum time for external precharge.

Figure 6.15(a) shows a read with an auto-precharge for a CAS\ latency of 3 and a burst length of 4.

The command to read bank 0 with auto-precharge is given on the same clock as the CAS\ command. After a time equal to the burst length, bank 0 is precharged. Then after a time tRP has elapsed, bank 0 can be activated.

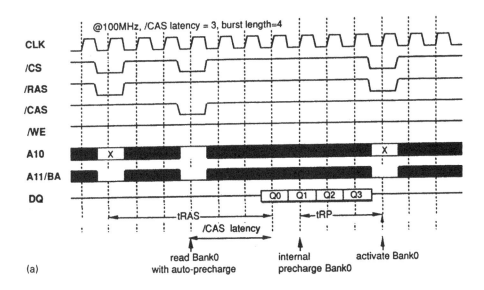

(a)

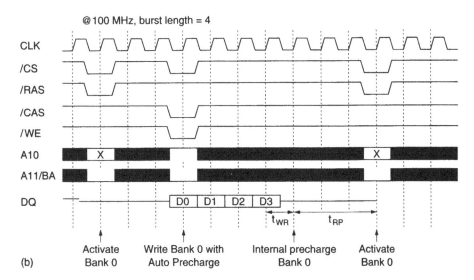

(b)

Figure 6.15   (a) SDRAM read cycle showing auto-precharge command timing (source: Mitsubishi [20]); (b) auto-precharge timing for a write

6.12.3   External Precharge Timing

A command to precharge the specified bank or all banks can be given during the read cycle. In this case, A11 is used to specify the bank to be precharged and A10 is used to indicate the precharge option. A10 = 0 means to precharge the bank specified by A11, and A10 = 1 means to precharge all banks.

An external precharge command cannot be given more than a CAS\ latency before the end of the burst read cycle without interrupting the burst. External precharge must obey the 2N rule.

For a CAS latency of 1, the minimum requirement is that the precharge commands coincide with output of the last data from a burst regardless of burst length.

The precharge requirement for CAS\ latency is illustrated in the timing diagram in Figure 6.16.

The timing diagram for a read in Figure 6.17 gives an example of external precharge using the same conditions as in Figure 6.15(a).

The figure shows the external precharge command coinciding with the second word of a read burst of 4. The precharge is complete after a time equal to the t_{RP} and bank 0 can then be activated. The burst is not interrupted since the precharge command is given no more than one CAS latency before the end of the burst.

External precharge for writes requires that the write burst be completed and time allowed for the write recovery (t_{WR}) of the sense amplifiers before a precharge may be signaled (in this case the time is one cycle). An external prepcharge command given during a burst write would interrupt the burst. The burst would be interrupted from the word of the burst before which the precharge command occurred.

An auto-precharge for a write would also cause the precharge command to be issued at time equal to the write recovery delay after the end of the write burst. This illustrated in Figure 6.15(b).

6.12.4 Write Latency

The write latency for the SDRAM is defined as the clock cycle difference between the clock where write command and column address are asserted and the clock where first data to be written is asserted. A write cycle with write latency of "0" is standard for the SDRAM I as shown in in Figure 6.18.

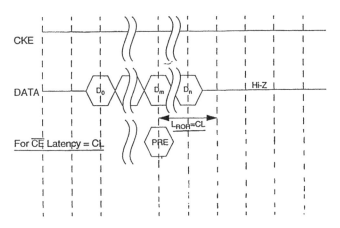

Note: Clock low-to-high transitions occur at the dotted lines

Figure 6.16 One CAS\ latency before the end of a read burst is the earliest a precharge command can be given without interrupting the burst (source: EIA JEDEC Standard 21C)

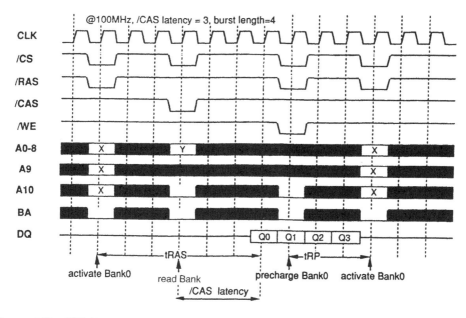

Figure 6.17 SDRAM read cycle showing external precharge command timing (source: adapted from Mitsubishi [20])

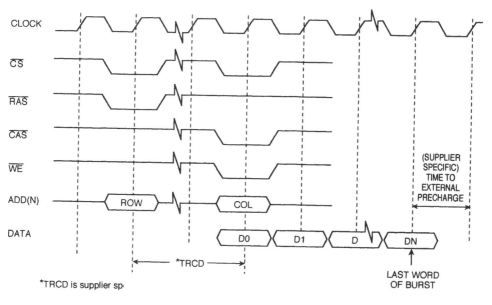

Figure 6.18 Timing diagram illustrating write latency (source: adapted from EIA JESD 21C [21])

6.12.5 DQM Latency for Reads

The DQM signal is the data mask for both reads and writes. During reads, DQM performs as a synchronous output enable. The DQM latency for reads is standardized as 2 and is the difference between the clock when DQM is asserted and the clock when the output bus has been forced to high Z. This is shown in the diagram in Figure 6.19 illustrating DQM for reads.

6.12.6 DQM Latency for Writes

During writes DQM performs the function of write data masking. The requirement for high speed operation and the synchronous nature of the device requires that the DQM latency be different for reads and for writes.

The DQM latency for writes is "0" and is defined as the difference between the clock when DQM is asserted and the clock when the write input data is inhibited as shown in Figure 6.20.

6.12.7 The 2–N Rule

The minimum column address to column address delay in page mode is standardized at two clock cycles regardless of frequency. This 2N rule is effective during bursts. That is, if the initial read or write command of a burst occured on an odd (even) clock cycle, the new column addresses must be presented on an odd (even) clock cycle while the burst is in progress.

The 2–N rule is also effective during interrupted bursts where the column address must, at a minimum, follow the 2–N rule while a read or write burst is in progress. After the burst is completed, the 2–N rule no longer applies and a new column address may be presented to the SDRAM on any clock cycle. The 2–N rule is in effect for a time "CAS latency + burst length" clock cycles for reads, and "write recovery + burst length" clock cycles for writes.

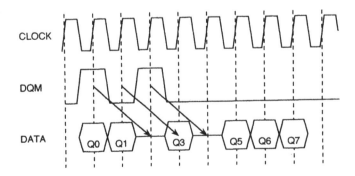

Figure 6.19 Timing diagram illustrating DQM for reads with read latency 2 (source: adapted from EIA JESD 21C [21])

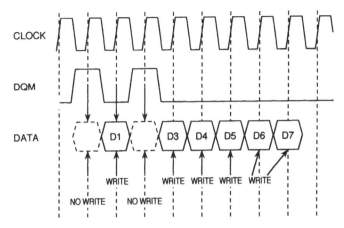

Figure 6.20 Timing diagram illustrating DQM for writes with write latency 0 (source: adapted from EIA JESD 21C [21])

This is illustrated in the timing diagram for CAS\ to CAS\ delay in Figure 6.21. While the 2–N rule is required of all JEDEC SDRAMs, not all JEDEC SDRAMs need the rule. Those with a 2-bit prefetch require it. Pipelined SDRAMs can do back-to-back CAS\ accesses during a burst and do not require it.

6.13 Refresh and Power Down on the JEDEC SDRAM

Refresh is needed on the synchronous DRAM since the memory array is still composed of dynamic cells which lose their charge without periodic refresh. The normal refresh mode is called auto refresh and is, internally, the usual DRAM "CAS before RAS" refresh.

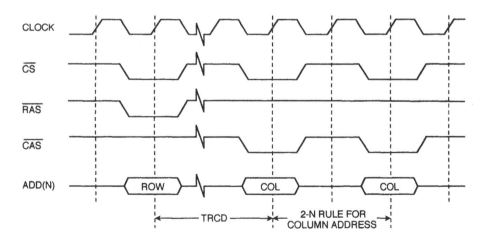

Figure 6.21 2–N rule for CAS address delay during a burst (source: adapted from EIA JESD 21C [21])

6.13.1   Auto Refresh Function

The auto refresh operation initiates a single refresh cycle that is completed internally by the SDRAM. An internal register supplies the refresh addresses for a mode that is internally similar to the CBR refresh on the standard DRAM.

Before performing an auto refresh, all banks on a multi-bank SDRAM must be precharged (i.e. closed). Precharge is accomplished by selecting an "idle" state (see the truth table in Figure 6.14).

Auto refresh then occurs when RAS\ and CAS\ are asserted on the same cycle. All banks automatically precharge at the end of the refresh cycle.

Since the SDRAM is underlaid with a standard DRAM, the minimum refresh time must be obeyed following an auto refresh before the next command is supplied to the device. The timing diagram in Figure 6.22 illustrates the auto refresh function.

6.13.2   Self Refresh

In addition, a "self refresh" mode has been defined for the SDRAM during which the part refreshes itself without outside intervention. Self refresh simplifies refresh control since it is completely controlled internal to the chip, and retains data in a low power mode.

In self refresh mode, the clock is disabled since the CKE pin remains low. All other inputs are disabled and inputs on any pins other than the CKE pin are ignored. During self refresh the power down mode is initiated with all CLK buffers suspended internally. Data retention power becomes comparable to a standard DRAM.

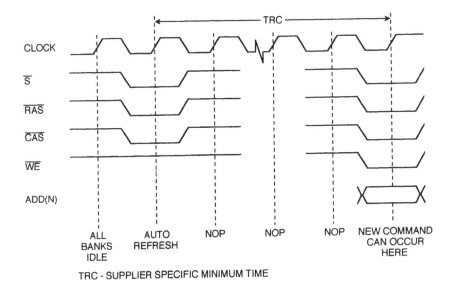

Figure 6.22   Illustration of auto refresh (source: adapted from EIA JESD 21C [21])

To enter self refresh mode the proper command from the truth table in Figure 6.14 is given (RAS\, CAS\, CS\, CKE = low, and WE\ = high). During self refresh, as long as CKE remains low all other signals can be at an arbitrary level.

A self refresh timing diagram is shown in Figure 6.23.

Self refresh begins by precharging all banks, then taking the CKE pin low. An autorefresh command is then given with CKE remaining low and on the next cycle the part enters self refresh and the clock can stop.

To exit self refresh mode, as shown in Figure 6.23, the clock (CLK) resumes first, then CKE goes high. After a recovery time of a RAS\ cycle (t_{RC}) any operation command can be given.

6.14   Power Down and Clock Enable

6.14.1   Clock Enable

The SDRAM can be placed in the power down mode by use of the clock enable (CKE) pin. The CKE pin governs the operation of the clock and as a result the "power down" and "suspend power" states.

A functional truth table for the clock enable function is shown in Figure 6.24.

Operations shown on this table include: self refresh entry and exit, auto refresh command, power down entry and exit, and clock suspend.

6.14.2   Power Down

The power down, or low power state, can only be entered using CKE if all internal banks of the device are precharged, that is, in the IDLE state as defined in the truth table in Figure 6.14 for device deselect. Once the SDRAM is in the IDLE state, the CKE pin can control the input buffers of the SDRAM.

Two commands for entering power down are shown in the CKE truth table in Figure 6.24. Both take the clock enable pin from high to low with all banks idle.

The timing diagram in Figure 6.25 illustrates the power down mode.

It is important to note that it is not possible to refresh the SDRAM during power down, so the minimum refresh specification must be observed. Since the minimum refresh timing is determined by the technology, it differs from supplier to supplier and must be specified for each part.

There are also two methods to exit power down: CKE can go high, or RAS\, CAS\ and WE\ can go high together. After exiting power down a minimum start-up time (T) must also be observed. T is also supplier specific and must be specified for each device. On the next rising clock after time T, a new command will be accepted.

6.14.3   Clock Suspend

The clock is suspended by taking CKE low and restarted by taking CKE high.

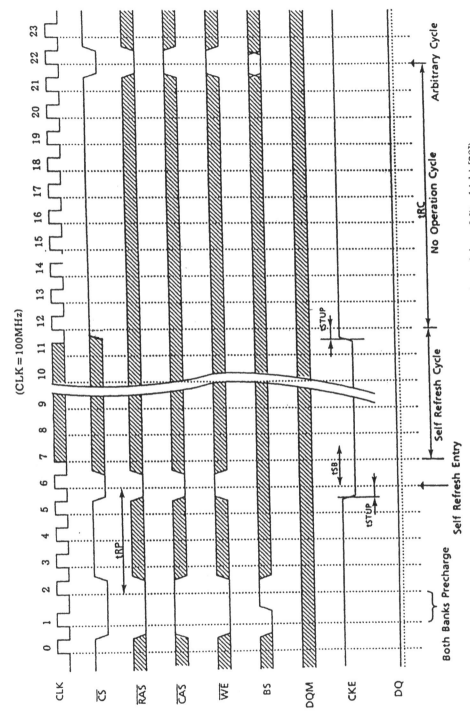

Figure 6.23   Self refresh timing diagram (source: adapted from Mitsubishi [20])

CURRENT STATE	OPERATION	CKE(N-1)	CKE(N)	CS\	RAS\	CAS\	WE\	ADDR
SELF-REFRESH	EXIT SELF REFRESH	L	H	H	X	X	X	X
	EXIT SELF REFRESH	L	H	L	H	H	H	X
	MAINTAIN SELF REFRESH	L	L	L	X	X	X	X
POWER DOWN	EXIT POWER DOWN	L	H	H	X	X	X	X
	EXIT POWER DOWN	L	H	L	H	H	H	X
	MAINTAIN LOW POWER MODE	L	L	L	L	L	X	X
ALL BANKS IDLE	ENTER POWER DOWN	H	L	H	X	X	X	X
	ENTER POWER DOWN	H	L	L	H	H	H	X
	ENTER SELF REFRESH	H	L	L	L	L	H	X
	AUTO REFRESH COMMAND	H	H	L	L	L	H	X
CLOCK SUSPEND	EXIT CLOCK SUSPEND	L	H	X	X	X	X	X
	MAINTAIN CLOCK SUSPEND	L	L	X	X	X	X	X
ANY	BEGIN CLOCK SUSPEND	H	L	X	X	X	X	X

Figure 6.24 Truth table for clock enable

In addition, if both of the internal banks are not in the IDLE state when CKE is asserted, the effect is to suspend the clock for two or more cycles as shown in the timing diagram in Figure 6.26. This figure illustrates that CKE taken low during a burst read, in this case on burst Q0, suspends the clock on the next cycle of the burst. The data for Q1 appears at the output buffer but is not clocked out while the clock is suspended. It is, in this case, only suspended for one cycle so that Q1 is clocked out with a one cycle suspension.

For the write suspend, also illustrated in Figure 6.26, the clock is suspended the cycle after the CKE is taken low during a write burst, and the data which would have been written in the next cycle is not written in this cycle but on the next. The clock

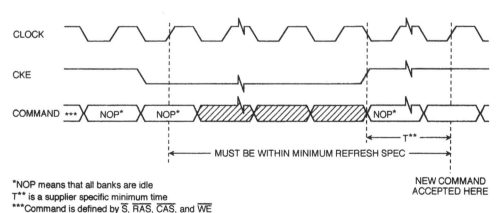

*NOP means that all banks are idle
T** is a supplier specific minimum time
***Command is defined by $\overline{S}$, $\overline{RAS}$, $\overline{CAS}$, and $\overline{WE}$

NEW COMMAND
ACCEPTED HERE

Figure 6.25 Power down mode timing diagram (source: adapted from EIA JESD 21C [21])

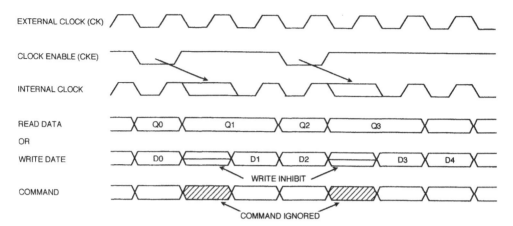

Figure 6.26 Timing diagram illustrating clock suspend (source: adapted from EIA JESD 21C [21])

suspend during write differs from a write mask operation where the masked data is never written. The suspended data is only delayed and does get written in on a later cycle.

6.15   A State Diagram for the JEDEC SDRAM

Since the SDRAM is completely controlled by truth table commands which put it into different operating states, its operation can be represented by a state diagram. While the complete state diagram of the multibank SDRAM is too complex to be illustrative, the usual operations can be represented in the simplified state diagram of the operation of the SDRAM shown in Figure 6.27.

6.16   Power On Sequence for the JEDEC SDRAM

A sequence for powering up the JEDEC SDRAM is recommended to ensure that the part powers up in a high impedance output state to avoid possible data contention in a multichip system where the chip might randomly power up at the same time that data is being driven out.

The recommended power-on sequence is shown in Table 6.3. If this sequence has been followed, the device should be in the IDLE state, that is, precharged and ready for normal operation.

For those planning to build or use an SDRAM, current, authoritative information on the JEDEC Standard Synchronous DRAM is available from the EIA/JEDEC office in Arlington, Virginia as the JESD 21C standards document for memories. It is important to check on the current standard since standards are updated on an ongoing basis to reflect the ongoing progress in the DRAM area.

Table 6.3 Recommended power-on sequence for JEDEC SDRAM

1. Apply power and start clock. Attempt to maintain a NOP condition at the inputs.
2. Maintain stable power, stable clock, and NOP input conditions for a minimum of 200 ms.
3. Issue precharge commands for all banks of the device.
4. Issue eight or more autorefresh commands.
5. Issue a mode register set command to initialize the mode register.

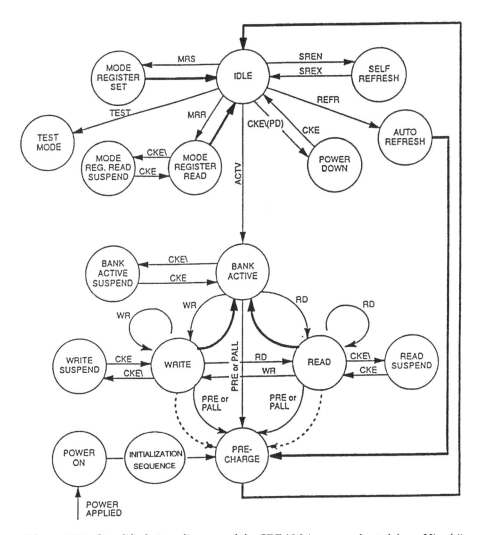

Figure 6.27 Simplified state diagram of the SDRAM (source: adapted from Hitachi)

6.17   Interface Options for the JEDEC SDRAM

The most common interface available on the 16M SDRAMs is the 3.3 V JEDEC Standard LVTTL interface. The early 64M SDRAMs also used the LVTTL interface including those complying with the PC100 specification commonly used in PCs. It is generally agreed that a low swing interface will be needed on the SDRAMs faster than 100 MHz. Common low swing interfaces used for faster DRAMs include the SSTL3 used for 3.3V SDRAMs, the SSTL2 used for 2.5V SDRAMs and for the SLDRAM, and the RTL used for the Rambus DRAMs.

The various standard low swing and differential interfaces potential are described in Chapter 8. It is also expected that a power supply voltage lower than 3.3 V will be used on future generations of SDRAMs.

6.18   64M, 128M, and 256M SDRAM I (SDR)

The basic SDRAM specification is called the SDRAM I or Single Data Rate (SDR) SDRAM. It has been used for the 16M through 256M SDRAM densities. The SDRAM II (DDR) specification will be discussed in a later section.

6.18.1   64M SDRAM I

The 64M SDRAMs are designed to be faster and wider than the first generation 16M SDRAMs.

A typical 150 MHz four-bank 64M SDRAM in 0.35 micron technology was described by Fujitsu in 1994 [14] with an area penalty of about 1.5 percent over the asynchronous DRAM. This part combines the benefits of a pipeline architecture with those of prefetch, picking up the best of both aspects of the earlier generation of parts. It includes the original LVTTL interface with the option for a low swing interface for higher speeds.

The column access path is divided into a three stage pipeline resulting in a three-clock-cycle read latency with a 6.5 ns clock cycle.

The array is designed to facilitate the prefetch operation with each of the four banks having two cell array blocks separated by a word decoder as shown in Figure 6.28. During a burst read or write operation the separated cell array blocks are both activated. The dataword from both sides of the word decoder is read or written simultaneously resulting in a two-bit prefetch.

The cell array is divided into even and odd addresses within each bank which eliminates the need for additional logic circuitry.

The first high volume usage of SDRAMs is expected to occur at the 64M density where a 2M×32 SDRAM running at 125 MHz could give a bandwidth of 500MB/sec.

The JEDEC Standard 64M SDRAM I is defined with either two or four internal independent banks. It tends to use a 3.3 V power supply and commonly used standard interfaces are the 3.3V LVTTL or the SSTL3 low swing interface which is used for higher speeds (generally above 100 MHz). Operational features are similar to those of the 16M SDRAM.

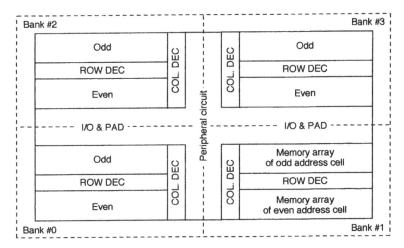

Figure 6.28   Chip architecture of a 64M SDRAM showing prefetch implementation (source: Y. Kodama [14] permission of IEEE)

A pinout for the 54-pin 0.8-mm lead TSOP Type II used for the 64M SDRAM I is shown in Figure 6.29. Pin assignment for the different output widths are shown including the ×4, ×8, and ×16 SDRAM.

Notice that the standard pinout has a power or ground pin on either side of each input/output pin. This is done to reduce the ground bounce phenomena at the higher speeds of the 64M DRAM. There are also two types of power and ground pins — V_{SS} and $V_{SS}Q$ and V_{DD} and $V_{DD}Q$. The basic power and ground, V_{DD} and V_{SS}, go to the array, while $V_{SS}Q$ and $V_{DD}Q$ power the I/Os. This maintains a quieter supply line to the DRAM array since the high speed interface can be isolated from the array power lines. A V_{REF} pin can be added on pin 40 in place of the no-connect (NC) for the reference voltage used by the SSTL interface. Other pins are similar to those used by the 16M SDRAM. V_{CC} is an older notation for 5V V_{DD}.

A typical 3.3V LVTTL 64M SDRAM I has specified clock frequencies of 83 MHz, 100 MHz, and 125 MHz. The four-bank operation is controlled by two bank address pins — BA0 and BA1. Burst length is programmable as 2, 4, or 8 bits, although many parts also offer a non-standard burst length of 1 bit, which is used during test.

The CAS\ latency is programmable at 2 or 3 clock cycles. This reflects an underlying t_{CAC} of 24 ns for a CAS\ latency of 3 clock cycles at 100 MHz to 125 MHz and 2 clock cycles at 83 MHz. The refresh is 4096 cycles in 64 ms.

A block diagram illustrating the four banks of the standard 64M SDRAM is shown in Figure 6.30. Typically, a wide internal bus takes two or four words at a time from any one bank on a wide internal bus to a prefetch register on the output. The individual words are then interleaved out at the specified datarate. This permits 125 MHz external speed to be achieved while running the internal array at a slower speed of 66 MHz or less.

The state diagram for this part is identical to that shown in Figure 6.27 for the 16M SDRAM. All legal states and transitions between these states are shown. From an idle state with all four banks precharged, the mode register can be set, the part can power down and enter self refresh, and an auto refresh can be done. If a row in one of the

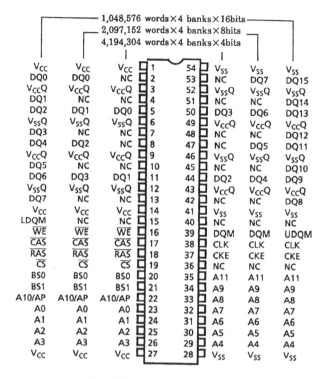

Figure 6.29 Pinout for the 64M SDRAM×4, ×8, ×16 in 54 pin TSOP II (source: Toshiba 64M SDRAM Datasheet

banks is opened using the "activate" command, then from this new row active state, the clock can be suspended, a word in the row can be read or written, the row can be closed and the bank precharged, Read with auto-precharge or write with auto-precharge are also options, and a read or write burst can be suspended for a cycle using the clock enable function.

Single bank read and write operations are the same as shown previously for the 16M SDRAM. A multiple bank burst read operation was shown in Figure 6.9 for the 16M SDRAM and is the same for the 64M SDRAM.

Figure 6.31 shows timing diagrams for auto-precharge operations with burst length of 4, CAS latency of 3 and clock frequency of 100 MHz.

Note that these timing diagrams no longer show the logic levels of the individual pins. They show only the clock, commands, and input/output operations. Since the SDRAM is effectively a digital state machine, all of the detailed timing signals are no longer necessary in the timing diagrams. These simplified timing diagrams are commonly used in datasheets for the SDRAM and, in addition, the full timing diagrams are also usually included for those interested in understanding the underlying DRAM operation.

Figure 6.31 (a) shows a read with auto-precharge. On the first clock, the ACT command is given, the row addresses are set up, and bank 1 is selected. After three clock cycles to allow sufficient time for t_{RCD}, the read command is given, the

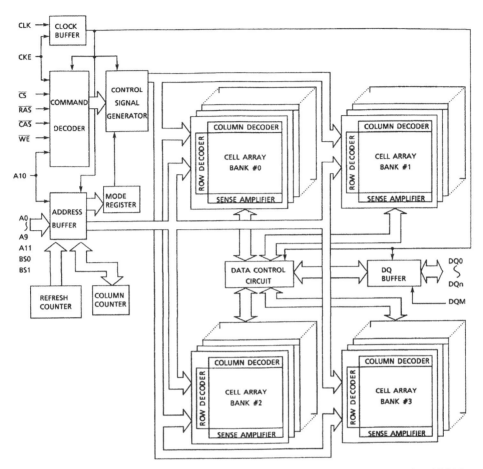

NOTE :   4,194,304 words×4 banks×4 configuration is 4096x1024x4 of cell array with the DQ pins numbered DQ0-3.
2,097,152 words×4 banks×8 configuration is 4096x512x8 of cell array with the DQ pins numbered DQ0-7.
1,048,576 words×4 banks×16 configuration is 4096x256x16 of cell array with the DQ pins numbered DQ0-15.

Figure 6.30   Block Diagram for the 64M SDRAM I (source: Toshiba 64M SDRAM I Datasheet [26])

column addresses are set up, and Address 10 is high, indicating an auto-precharge is to occur after the burst is complete. Three cycles after the read command (the CAS latency time), the first word of the burst appears on the output. Four cycles after the read command was given (the burst length), the auto-precharge begins. The bank is finished precharging after a time (t_{RP}) which is equal to the CAS latency and another ACT command can be given with another row address being set up.

If an external precharge had been used instead for this read cycle. A10 would have been left low during the read command and the external precharge command could be given as early as three cycles (the CAS latency) before the end of the burst to precharge the bank at the end of the burst without interrupting the burst.

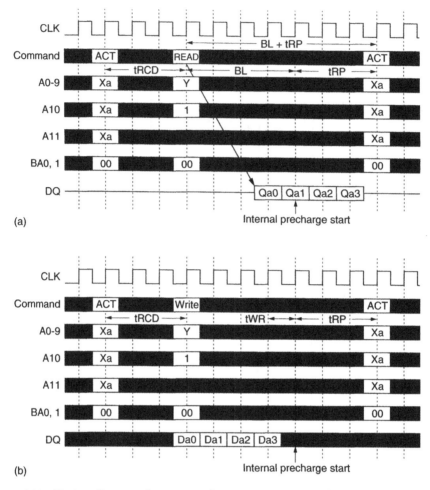

(a)

(b)

Figure 6.31 Timing diagrams for auto-precharge operation a. read with auto-precharge (BL=4, CL=3); b. WRITE with Auto-Precharge (BL=4) (source: Mitsubishi 64M SDRAM I Datasheet[24])

Figure 6.31 (b) shows a write with auto-precharge. On the first clock, the ACT command is given, the row addresses are set up, and bank 1 is selected. After three cycles for t_{RCD} the write command is given, the column addresses are set up, and A10 is high to signal an auto-precharge. Since the write latency is equal to zero, the first word of the data burst is also clocked in on this cycle. The auto-precharge cannot begin until all the words of the burst are clocked in and a cycle allowed for the sense amplifiers to stabilize — the write recovery time (t_{WR}). After a three-cycle precharge (t_{RP}), another ACT command can be given and another row opened.

Figure 6.32 shows a read and a write cycle using the DQM function. Recall the DQM latency for a write is zero for the SDRAM I. The DQM is, therefore, shown high on the second word of the write burst and the data word in is masked on that same cycle. For a read, the DQM latency is equal to two so the word that is masked is the

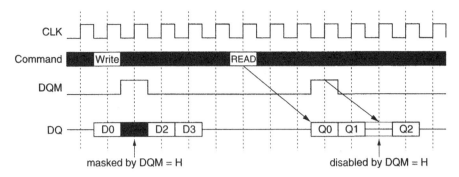

Figure 6.32   DQM masking function During a read and a write cycle. (source: Mitsibushi 64M SDRAM I Datasheet [24])

third word of the read burst even though the DQM signal is high on the first word of the burst. Note that the word masked by the DQM is lost in both the read and write case, with the burst continuing on the following word of the burst.

An illustration of suspending a burst cycle without losing the word of data is shown in Figure 6.33. This is achieved with the Clock Enable (CKE) function. CKE low during a burst suspends the function for the next cycle. In the case of a write, the data word is masked on the next cycle after the CKE is low and is then written on the following cycle. In the case of a read, in the cycle after the CKE low is signaled, the word of the burst appears on the output and is held until it is clocked out on the next cycle.

The power down and refresh functions are identical to those of the 16M SDRAM I. If all banks are precharged (idle) when the CKE goes low, then the chip goes into standby power-down and if a bank is active when the CKE goes low, the chip goes into active power down. If the chip is powered down without activating the self refresh feature, then it must be powered up again within the specified refresh time to avoid losing data.

The SDRAM can be put into self refresh mode. If all banks are precharged when the CKE goes low, and the self refresh command is given on the clock cycle after the chip enters standby power down, the chip refreshes itself until the CKE is taken high

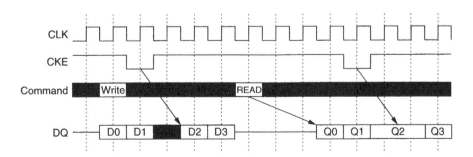

Figure 6.33   Suspending a burst cycle using Clock Enable (CKE) (source: Mitsibushi 64M SDRAM I datasheet [24])

and one refresh time (t_{RC}) has elapsed. The requirement to wait for the additional cycle of refresh accommodates the case where a refresh cycle has just begun when the CKE high command is given.

Auto refresh can be signalled at any time that the banks are precharged by giving an auto refresh command and waiting a minimum refresh cycle time (t_{RC}).

Among functions that are permitted are: read burst interrupted by another read burst and write burst interrupted by another write burst. These operations are shown in Figure 6.34(a) and (b).

In Figure 6.34(a) it can be seen that a read can interrupt a read in the same bank or in different banks on any cycle. The effect is to interrupt the burst from the first read with the burst from the second read. Similarly in Figure 6.34(b) we see that a write can interrupt a write in any bank on any cycle. Neither a read with auto-precharge nor a write with auto-precharge can be interrupted.

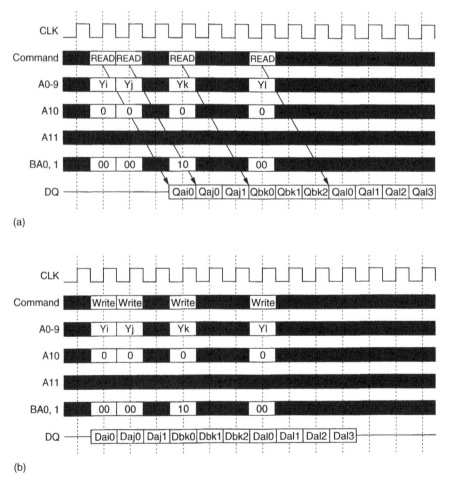

(a)

(b)

Figure 6.34   Example I of read and write interrupt operations: (a) read interrupted by a read (BL=4, CL=3); (b) write interupted by a write (BL=4) (source: Mitsubishi 64M SDRAM I Datasheet [24])

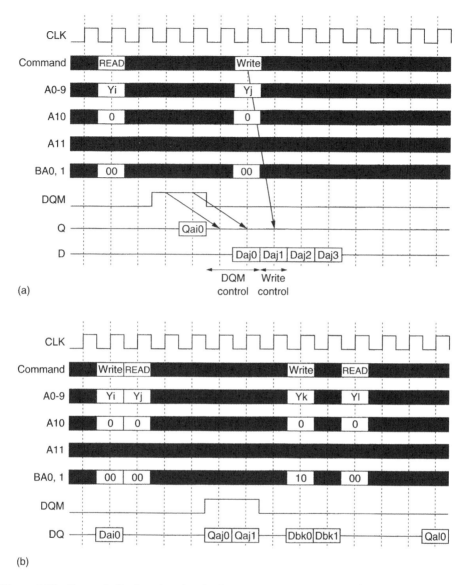

Figure 6.35   Example II of read and write interrupt operations. a. read interrupted by a write (BL=4, CL=3); b. write interrupted by a read (BL=4, CL=3) (source: Mitsibushi 64M SDRAM I Datasheet [24])

The case of read interrupted by write and write interrupted by read is shown in Figure 6.35(a) and (b).

If a read is to be interrupted by a write, it is necessary to turn the bus around. This case is shown in Figure 6.35(a). The DQM function is used to take control of the bus from the read. In this figure, it is intended to interrupt the read burst after the first word of the burst. Since the DQM latency for a read is two cycles, it is necessary to signal the DQM high command a cycle before the first word of the data burst appears. This gives the DQM control of the bus on the clock cycle for the second word of the read burst and masks this word. Since the DQM is held high for two cycles, both the second and third word of the burst are masked. On the cycle for the third word of the read burst, the write command can be given and the first word of the write burst be written in on this cycle. The write command takes control of the bus on the second word of the write cycle.

A write command can be followed by a read command on the next cycle as shown in Figure 6.35(b). The effect is to mask the word of the write burst on the cycle where the read command is given and terminate the write burst. The figure also shows the read cycle being interrupted by another write cycle using DQM to gain control of the databus.

It is possible to achieve a gapless burst by reading from one bank while another bank is being precharged. A multi-bank interleaved read burst is shown in Figure 6.36.

Bank B is activated after the read command for bank A is given and the read command is given to bank B as soon as t_{RCD} is satisfied and during the bank A burst. The result is to have a gapless burst of 8 bits. In this case, external precharge is used to precharge bank A while the burst is being read from bank B. Following the precharge, another burst can be read from bank A, but this second burst from bank A will not be gapless with the second burst from bank B.

If the burst length had been 8, continuous gapless burst could be achieved with two banks. Also if the burst length is 4 and four banks are interleaved, then a continuous gapless burst could also be achieved.

A gapless multibank interleaved write burst is shown in Figure 6.37. A minimum of "t_{RCD}" after bank activation, the write command can be given. Since write latency is equal to zero, the write burst begins on the same cycle as the write command. An activate to a different bank can be given on the following cycle and a new write command to the second bank can be given following completion of the burst from the first cycle.

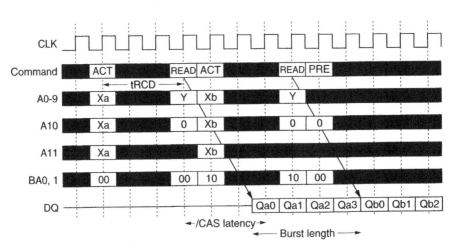

Figure 6.36 Multibank interleaved read burst (BL=4, CL=3) (source: Mitsubishi 28M SDRAM I Datasheet [24])

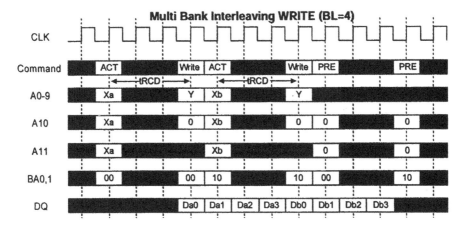

Figure 6.37 Multibank interleaved write burst (source: Mitsubishi 128M SDRAM Datasheet)

Page mode is also possible using the SDRAM. Recall page mode involves accessing more than one word from a given page before closing that page and opening another. Since no banks need to be closed, it is not necessary to allow time for precharge. In this case, a full page of gapless burst is possible. Figure 6.38 shows a timing diagram for an SDRAM I page mode burst read.

As shown in the figure, page mode involves leaving a bank open and clocking in multiple read commands for addresses on the open sense amplifiers. The function is completely analogous to page mode for the asynchronous DRAM.

A new optional operating mode for the SDRAM I has been added — the burst terminate mode. A "read with burst terminate" timing diagram is shown in Figure 6.39(a) and 6.39(b) and a "write with burst terminate" is shown in Figure 6.39(c). In all cases the assumption is made that the burst length is four or longer.

As shown in the illustration, in a read cycle the burst terminate command has the same latency as the CAS latency. In a write cycle the burst terminate command has latency equal to zero the same as the write latency. Burst terminate is not a valid command during a read or write with auto-precharge.

Many of the 64M SDRAM I units shipping into the PC application will be in "PC100" specified SDRAM modules. This specification, originally defined by Intel, requires 6 ns of access time from clock (t_{AC}) at 100 MHz clock speed with 50 pF capacitive loading.

The definition of access time from clock as well as command set-up and hold timing can be seen in Figure 6.40, which shows a full timing diagram of a read without auto-precharge operation.

While SDRAM operation eliminates many of the set-up and hold timings required for the asynchronous DRAM, the set-up and hold timings for commands remain. In addition, the data out on a read must be stable some time before the rising clock edge which clocks it out. The parameter that is measured is the access time from the previous rising clock edge defined as "access time from clock"(t_{AC}).

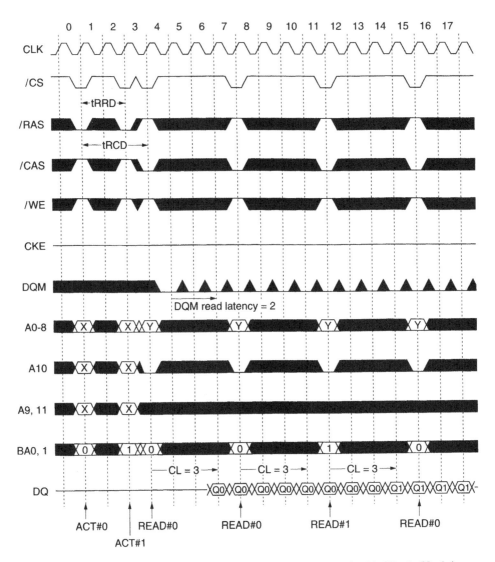

Figure 6.38 Timing diagram for SDRAM I page mode burst read with BL=4, CL=3 (source: Mitsubishi 64M SDRAM I Datasheet [24])

6.18.2 128M and 256M SDRAM I

After the 64M density, many systems did not require as much memory as offered by a 256M DRAM. As a result, a 128 Mb density was offered. For example, the average size of main memory in 1998 was 64MB and the PCI or PC100 bus in a PC was 64 bits wide. A 64MB system could, therefore, be made with 8 64Mb DRAMs organized 8M×8 on the 64-bit bus. With a 128M DRAM configured as 8M×16, a cost reduction could be achieved in this system by replacing the 8 64M DRAMs with 4 128M

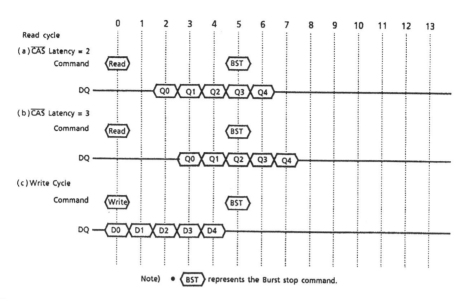

Note) • ⟨BST⟩ represents the Burst stop command.

Figure 6.39 Example of burst terminate commands for read and write. (a) read with burst stop (burst length greater than 5, CAS\latency = 2); (b) read with burst stop (burst length greater than 5, CAS\latency = 3); (c) write with burst stop (burst length greater than 5) (source: Toshiba 64M SDRAM Datasheet)

DRAMs as long as the price of the 128M DRAMs was less than two times that of the 64M DRAMs. Alternatively, a density upgrade to 128MB could be achieved by replacing the 8 64Ms with 8 128Ms configured 16M×8.

The 128M and 256M SDRAMs could be designed with ×4, ×8, and ×16 I/O widths on the same chip. A pin out showing the compatibility between the three different widths for the 128M is shown in Figure 6.41. The 256M adds an additional address "A12" on pin 36.

A block diagram for this 128M four-bank chip is shown in Figure 6.42. Each of the four banks is organized ×8. The arrays are organized 4096 rows by 1024 columns resulting in a 4K refresh operation.

Other than the four-bank operation and the size of the arrays, there is no significant difference between the 16M, 64M, 128M, and 256M SDRAM architecture.

The mode register is shown in Figure 6.43. Burst lengths of 1, 2, 4, and 8 are permitted. As before, the burst length of one is used only for test. CAS\ latency of 2 and 3 is permitted.

6.19   SDRAM II — Double Data Rate SDRAM

The inherent speed of the SDRAM I architecture is limited to about 200 MHz. In order to have a higher speed SDRAM, the industry fell back on the older architectural concept of a multiple bit prefetch. Prefetch was used in many of the first

Command Input Timing

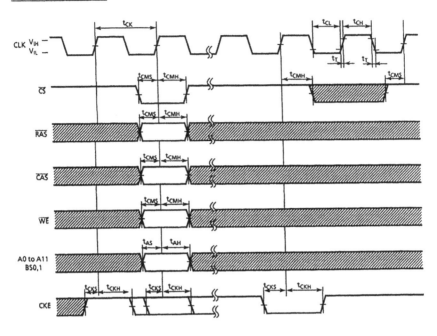

Read Timing

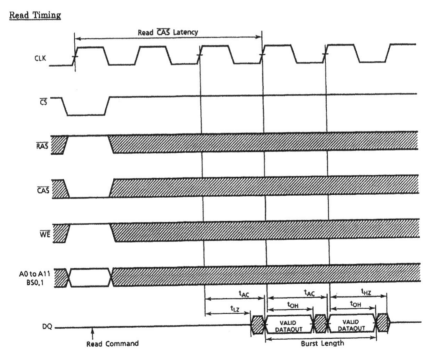

Figure 6.40 Timing diagram illustrating: (a) Command set-up and hold timings; (b) read command showing access time from clock (t_{AC}) (source: Toshiba 64M SDRAM Datasheet)

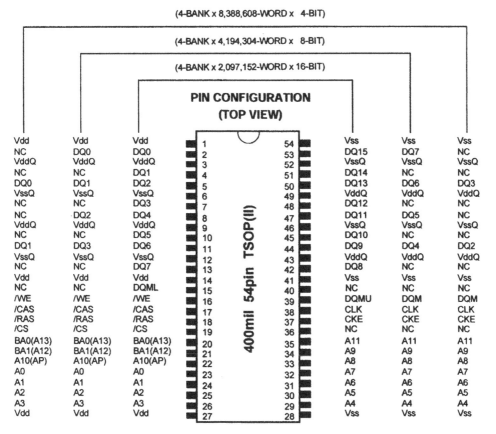

Figure 6.41 Pin assignment for 128M SDR SDRAM ×4, ×8, and ×16 width in 54 pin TSOPII (source: Mitsubishi 128M SDRAM Datasheet [25])

generation SDRAMs for the purpose of being able to run the array at a slower speed than the I/O. The concern at that time was about disturb problems in the DRAM array from multiple bank operation and from the noise of the higher speed interface. In this early case for the SDRAM I, a two-bit prefetch was used and the array was run at half the speed of the I/O. This led to the "2-*n*" rule which stated that column addresses during a burst could only occur on even clock cycles.

The SDRAM II, also called the Double Data Rate (DDR) SDRAM, uses a two-bit prefetch buffer on the output. The array is run at maximum speed and the two prefetched words are interleaved out at twice the array speed. This means that for each column address there are always two words of data fetched on a wide internal data bus. Since a column address can be given on every clock cycle, there are two words of data in or out on each clock cycle.

There are three basic external differences between the SDRAM I and the SDRAM II (DDR). The first is that the data runs at a rate that is doubled from the system clock. The second is that the write latency is one, and the third is that a differential clock is used instead of a single clock.

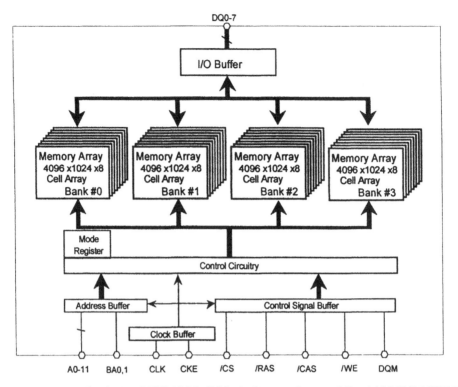

Figure 6.42 Four bank 128M SDRAM (×8) block diagram (source: Mitsubishi 128M SDRAM Datasheet [25])

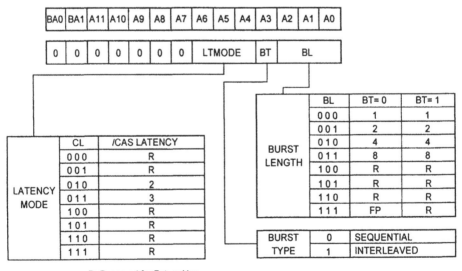

Figure 6.43 Mode register operation for typical 128M SDR SDRAM (source: Mitsubishi 128M SDRAM Datasheet [25])

The CAS latency for reads also needs to be modified to accommodate half cycle latencies. A comparison of the CAS latencies supported by the SDRAM I and II are shown in Table 6.4; numbers in parentheses are optional.

Examples of a 128M DDR SDRAM in 66 pin TSOP and in 60 pin CSP are shown in Figures 6.44a and 644b. New pins are the DQS or Datastrobe pin, and the CK\.

The wider internal bus and two-bit output buffer can be seen in the block diagram shown in Figure 6.45. The 8 bits are prefetched from the DRAM array to the 2-bit column prefetch register, the two 4-bit words are then fed from the prefetch register to the output at double speed.

In the block diagram other new internal features include the delay lock loop (DLL) and bidirectional datastrobe.

The DLL is added to shift the output data in time such that the output data is nominally aligned with the input clock of the DRAM. It aligns the datastrobe edge to the input clock edge and ensures that all devices operating at the same frequency in a system have the same timing for the data out relative to the clock, regardless of differences in density and technology. The DLL will remain locked to the input clock as long as the input clock frequency remains within the valid operating range for DLL operation. The DLL can be disabled to save power or to operate at frequencies lower than the minimum specified range for DLL operation.

The datastrobe pin, DQS, is a bidirectional datastrobe which toggles when the data appears on the DQ pins. The datastrobe increases system performance by providing a good clock edge to the high speed data. The datastrobe for a read is generated on the chip.

The mode register and mode register set timing for the DDR SDRAM are shown in Figure 6.46(a). These are similar to the SDR SDRAM except that the mode register bits are expanded to include both a BS0 and BS1 field. BS0 determines the use of an extended mode register. The extended mode register determines whether the delay lock loop (DLL) is enabled or disabled. The DLL is expected to be disabled primarily during test. The extended mode register and set timing is shown in Figure 6.46(b).

A basic DDR read timing cycle is shown in Figure 6.47. The datastrobe goes from high impedance to a logic low level one clock cycle before the data from a read burst appears on the DQ pins. This is referred to as the datastrobe "preamble". At the end

Table 6.4 CAS\ read latencies for SDRAM I and II

CL	SDRAM latency	DDR SDRAM latency
000	R*	R*
001	R*	R*
010	2	2
011	3	(3)
100	(4)	R*
101	R*	(1.5)
110	R*	2.5
111	R*	(3.5)

*Reserved for future use

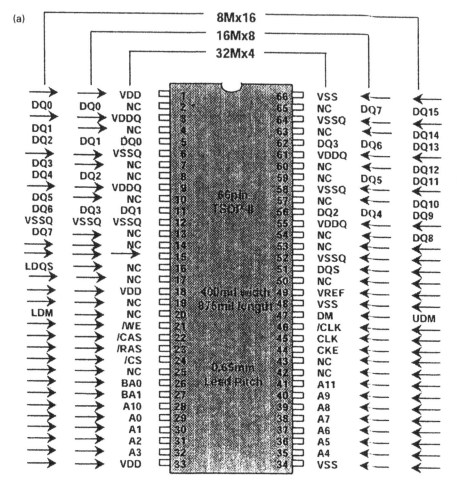

Figure 6.44   Pinouts for 128M DDR SDRAM. (a) In 66 pin TSOPII; (b) in 60 pin CSP (source: Toshiba 128M DDR SDRAM [26])

of a read burst, if there is no subsequent read burst, the datastrobe transitions from low to high. This is referred to as the datastrobe "postamble".

Set-up and hold times are specified for address, control and command signals that are sampled on the rising edge of the clock (CK). These input set-up time (t_{IS}) and input hold time (t_{IH}) minimum values are shown.

The skew between the datastrobe and the input clock (t_{DQSCK}) and the skew between the data out and the input clock (t_{AC}) are also shown. It is intended that the edges of the data out pins (DQ) and the datastrobe (DQS) during a read are nominally coincident with the edges of the input clock.

The minimum time that the output data is valid (t_{DV}) is important for the memory controller. This output data valid time is determined by the minimum clock cycle minus variations in data access and hold time due to DLL jitter and power supply noise.

Figure 6.44   *Continued*

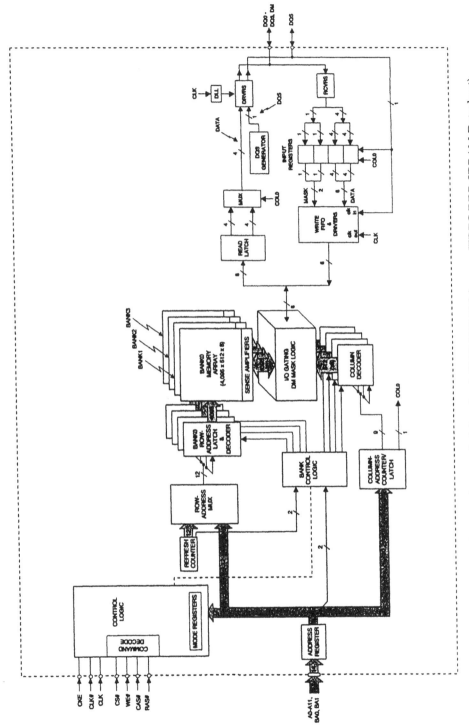

Figure 6.45 Block diagram for a typical four bank DDR SDRAM (source: JEDEC DDR SDRAM Datasheet)

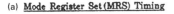

(a) Mode Register Set (MRS) Timing

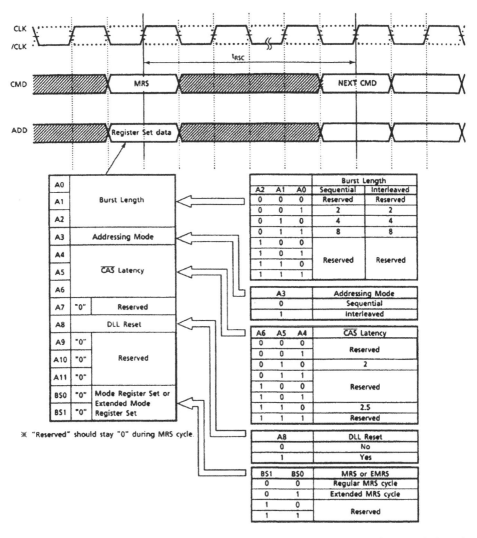

A0		Burst Length
A1		
A2		
A3	Addressing Mode
A4		CAS Latency
A5		
A6		
A7	"0"	Reserved
A8		DLL Reset
A9	"0"	
A10	"0"	Reserved
A11	"0"	
BS0	"0"	Mode Register Set or Extended Mode
BS1	"0"	Register Set

※ "Reserved" should stay "0" during MRS cycle.

			Burst Length	
A2	A1	A0	Sequential	Interleaved
0	0	0	Reserved	Reserved
0	0	1	2	2
0	1	0	4	4
0	1	1	8	8
1	0	0		
1	0	1	Reserved	Reserved
1	1	0		
1	1	1		

A3	Addressing Mode
0	Sequential
1	Interleaved

A6	A5	A4	CAS Latency
0	0	0	Reserved
0	0	1	
0	1	0	2
0	1	1	
1	0	0	Reserved
1	0	1	
1	1	0	2.5
1	1	1	Reserved

A8	DLL Reset
0	No
1	Yes

BS1	BS0	MRS or EMRS
0	0	Regular MRS cycle
0	1	Extended MRS cycle
1	0	Reserved
1	1	

Figure 6.46   DDR SDRAM. (a) Mode register and mode register set timing; (b) extended mode register and set timing (source: Toshiba 128M DDR SDRAM)

A timing diagram for a burst write operation for a DDR SDRAM is shown in Figure 6.47(a). During a write operation, the memory controller is required to provide the datastrobe. The transition of the datastrobe from high to low impedance nominally occurs on the falling edge of the clock after the write command has been received by the device. The datastrobe must be driven high nominally one clock after the write command has been registered. The datastrobe for a write operation is approximately centered within the data valid window. A data set-up time (t_{QDQSS}) and hold time (t_{QDQSH}) relative to the datastrobe are specified as shown in Figure

(b) Extended Mode Register Set (EMRS) Timing

AO	DLL Switch
A1	"0"
A2	"0"
A3	"0"
A4	"0"
A5	"0"
A6	"0"
A7	"0"
A8	"0"
A9	"0"
A10	"0"
A11	"0"
BS0	"1"
BS1	"0"

Reserved

Mode Register Set or Extended Mode Register Set

AO	DLL Switch
0	DLL Enable
1	DLL Disable

BS1	BS0	MRS or EMRS
0	0	Regular MRS cycle
0	1	Extended MRS cycle
1	0	Reserved
1	1	

※ "Reserved" should stay "0" during EMRS cycle.

Figure 6.46 *continued*

6.47(b). The intent is to minimize the skew between the data or data mask and the data strobe at the device.

This timing diagram shows that the write latency for the DDR SDRAM is 1. That is, the write command is given one clock cycle before the first word of the data burst is strobed in.

For a write operation interrupting a read operation on the DDR SDRAM, the burst terminate operation is used to stop the read burst and take the DQ bus to high impedance before the write data appears. Thus the burst terminate, which was optional on the SDRAM I, is mandatory on the DDR SDRAM. A write interrupting a read burst is shown in Figure 6.48. It is understood that a write command cannot interrupt a read with auto-precharge.

Since the burst terminate command is now used to turn the bus around for a write interrupting a read, the data mask function is no longer used for output masking. It,

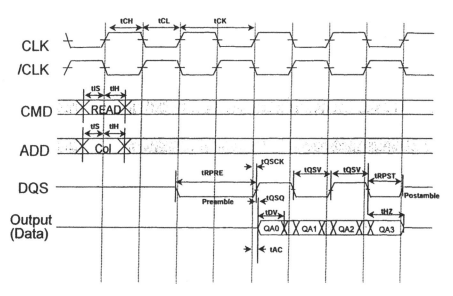

Figure 6.47a DDR read burst cycle timing (source: Toshiba 128M DDR SDRAM Datasheet [26])

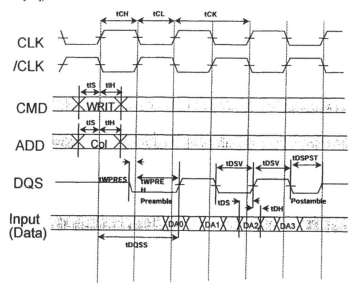

Figure 6.47b Timing diagram for a burst write operation for a DDr SDRAM (source: Toshiba 128M DDR SDRAM Datasheet [28])

therefore, becomes a DM function on the SDRAM II (DDR) rather than a DQM as on the SDRAM I.

The data mask is applied at the same time as the input data that it will mask. It is sampled on each edge of the datastrobe during write cycles and must satisfy the same set-up and hold conditions as the input data.

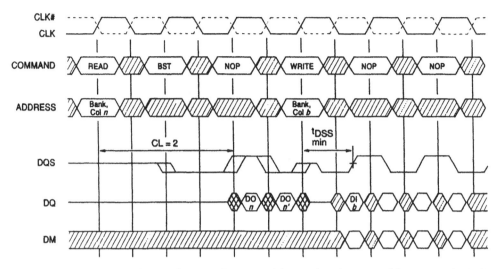

Figure 6.48 Example of write interrupting a read burst showing use of burst terminate to tristate the bus. CAS\latency = 2, burst length = 4 (source: JEDEC DDR SDRAM Datasheet)

For the case of a read interrupting a write, the data mask (DM) pin must be used to mask from the last two input data words before the read operation to one word after the read, as shown in Figure 6.49. Any subsequent words in the write burst are masked by the read. The latency for the DM continues to be zero as in the SDRAM I. The data mask (DM) is sampled on each edge of the datastrobe.

In Figure 6.49, the write command is given on the first clock and the first bit of data would be written in on the next clock cycle due to the one cycle write latency. The first four words of the data burst are written in, but the fifth and eighth are not written since the DM mask is active.

The read command is given on the seventh clock cycle and takes control of the bus on the nineth clock cycle tristating the DQ bus. At this time the driving chip releases the DQ and DQS buses in time to turn the buses around for the read. This allows the full cycle preamble for the datastrobe (DQS) for reads to occur before the read data appears.

6.20   Trends in SDRAM Characteristics

Some properties of the SDRAMs evolve with each generation. The number of banks went from two banks for the 16M SDRAM to four banks for the 64M SDRAM. The number of banks thus far has remained at four for the 256M SDRAM. The reason is that the advantage gained from increasing the number of banks is in permitting a gapless burst. Four banks continues to permit a gapless burst for the 256M, so it has not been necessary to add the complexity and cost of going to eight banks.

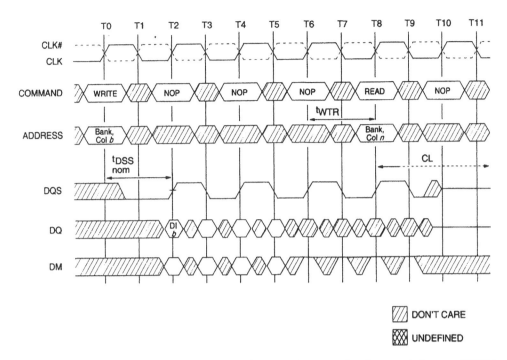

Figure 6.49 Example of read interrupting a burst write using the data mask (source: JEDEC DDR SDRAM Datasheet)

The speed of the DRAMs has historically increased with each generation since each density generation has been manufactured in a tighter technology and the tighter technologies tend to run faster. The progress of DRAM manufacturing technology has permitted a four-fold density improvement every three years; however, an increasing number of systems do not require the highest density DRAMs manufacturable. As a result, lower density DRAMS are being run in the latest manufacturing technology. Since these lower density DRAMs are smaller, they tend to be faster. The phenomenon seen now, therefore, is that a single density, such as 64M, is being run at faster speeds over time as the product is scaled to the latest technology.

Different architectures are used for faster speeds for the SDRAMs in an effort to match the speed requirements of the processors being used in the systems. This trend is plotted in Figure 6.50 which compares processor and DRAM speed trends by access mode.

6.21 Cache DRAMs

6.21.1 Introduction

Another innovation in DRAM speed has been the cache DRAM. At least three different types of cache DRAMs have appeared from major DRAM vendors. These are described in this section.

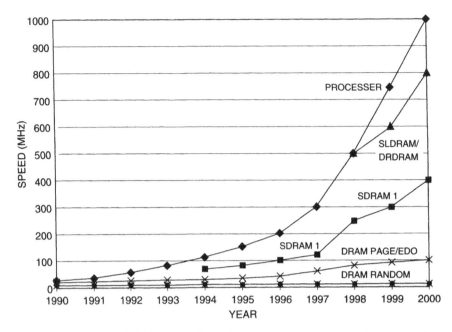

Figure 6.50 Processor and DRAM speed trends by access mode (MHz) (source: various datasheets; projections: Memory Strategies International)

A memory hierarchy using fast SRAMs as a "cache" between the processor and the main memory of a computer has been used for years to improve the datarate to the processor. The SRAMs, which are faster but more expensive than DRAMs, hold a small amount of frequently requested data between the processor and the DRAM.

The cache system works by the processor presenting addresses to the cache controller which determines if that address is available in the cache. If so, then the data is read from the fast SRAMs in the cache. If that address is not in the cache then a "miss" occurs, and the data is read from the slower DRAMs in main memory. A more complete description of this process can be found in Chapter 4.

Using both fast SRAM and slower DRAM is cost effective in a large computer since many low priced standard DRAMs and SRAMs can be configured by a single set of external logic chips into the cache hierarchy configuration.

In a smaller system, it is possible that an integration of the DRAMs, SRAMs and logic can produce a chip that is more efficient in terms of such factors as speed, power, package count, and PC board space. Examples of possibilities for this type of integration are shown in Figure 6.51.

The basic cache hierarchy is shown in Figure 6.51(a). A small level 1 cache SRAM is shown integrated on the processor chip in Figure 6.51(b). Most modern processors include an L1 cache. A level 2 standalone SRAM cache is commonly added for a faster system as shown in Figure 6.51(c). Standalone SRAM cache configurations were discussed in Chapter 4. A slower system might eliminate the L2 cache altogether, particularly if one of the high bandwidth DRAMs is used for the main memory.

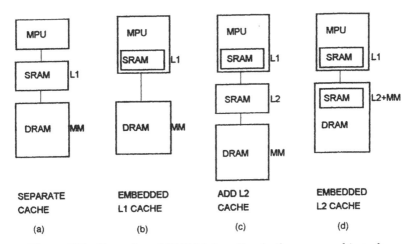

Figure 6.51 Examples of SRAM integration in the memory hierarchy

Another possibility, shown in Figure 6.51(d), is to integrate the L2 cache SRAM onto the DRAM chip to provide the required bandwidth to the processor directly from the main memory bank. The DRAM in this case is referred to as a cache DRAM. With cache DRAMs a higher bandwidth is achieved between the DRAM and the SRAM by eliminating the external interface between the two. The chip count is also reduced. This configuration would tend to be selected for a smaller system since the cost of each DRAM chip is higher than that of a standard DRAM. Beyond a certain size it would be more cost-effective to buy the standard components.

Manufacturers who have introduced or described such cache DRAMs include Mitsubishi, Oki, and Ramtron now called "Enhanced Memory Systems, Inc."

The cache DRAM introduced by Mitsubishi has separate SRAM and DRAM blocks on a single chip which are individually accessible and externally controllable. The cache DRAM from Oki uses the DRAM's sense amplifier banks as a cache. The Ramtron EDRAM uses dedicated registers at the opposite end of the bit lines from the sense amplifier. The separate banks of SRAMs and DRAMs are visible on the Mitsubishi chip while the closely tied registers of the Ramtron cache DRAM can not be distinquished from the DRAM array.

Cache DRAMs, as well as video DRAMs, use their wide internal buses to fill on-chip caches. Because of the direct link between the DRAM and the cache storage, a much higher data transfer rate is obtained than is possible with an off-chip cache.

6.21.2 The Enhanced DRAM

The early cache DRAM from Ramtron, the "Enhanced DRAM" or EDRAM, is the only one of the new generation of fast DRAMs that is asynchronous. As such it provides a link between the evolutionary fast asynchronous DRAMs covered in the previous chapter and the architecture-modified DRAMs discussed here.

The EDRAM is used primarily in unified memory in high end laptop computers to eliminate the L2 cache. This application requires nearly the speed of the desktop system in a smaller form factor to provide portability. The EDRAM permits simple and transparent unification of the main memory and the L2 cache with the ease of system design of dropping in a fast DRAM.

The SRAM, the DRAM and a comparator are combined on chip so the operation of the internal cache is hidden from the user. The SRAM is in the form of row registers closely coupled to the DRAM banks.

This is illustrated in the functional block diagram in Figure 6.52 of the 512K×8 EDRAM showing the four-bank cache architecture.

The 512K×8 enhanced DRAM uses 8K bits of SRAM direct mapped cache. The cache is distributed as 2K bits of row registers tightly coupled to each of four active DRAM pages on chip. 2K bits are available during a given cache read with 35 ns access to any page.

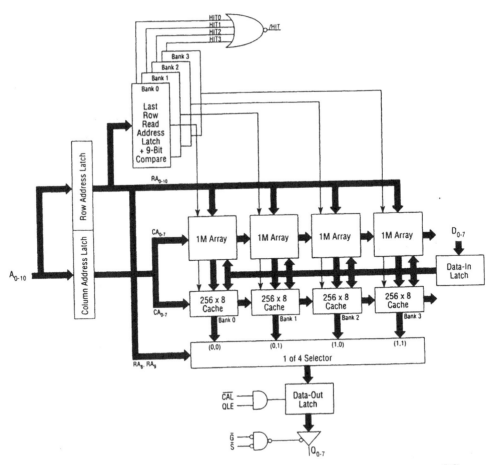

Figure 6.52   Functional block diagram of 512K×8 EDRAM (source: Ramtron [8])

The internal transfer between DRAM and SRAM is 256 bytes wide which permits fast cache fill. An output latch enable is provided to permit either fast page or Hyperpage (EDO) mode to be used.

Since the part externally operates like a standard DRAM, it can be operated with commercial DRAM controllers.

The 512K×8 EDRAM has four independent DRAM memory banks each with its own 256×8-bit SRAM row register. Memory reads always occur from the cache row register of one of the banks specified by row address bits A8 and A9 which are the bank select pins.

To keep the DRAM main memory and the cache coherent, a write-through cache technique is used with on-chip comparators. For a cache hit, data is written to the DRAM through the cache. For a miss, it is written directly to the array. Cache hit access time and cycle time is the same for both reads and writes. Cache miss access time and cycle time are longer since a new row in the DRAM must be accessed.

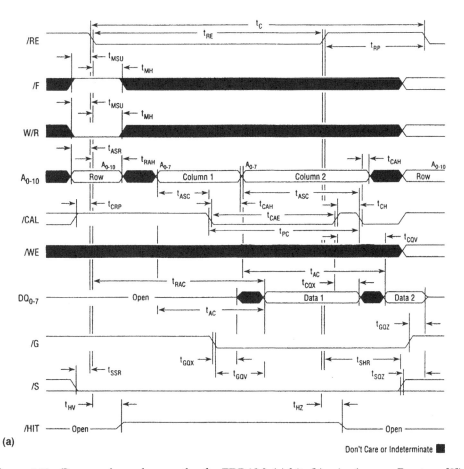

Figure 6.53 Page mode read access for the EDRAM: (a) hit; (b) miss (source: Ramtron [8])

A hit pin indicates the status of on-chip page hit or miss comparators which consist of a nine-bit row address latch. These latches together with the last row read address is reloaded on each active read miss cycle. Read or write burst cycle time is fast and a shorter output enable access time allows fast interleaving of the SRAM banks.

If the internal comparator indicates that the current requested row address matches that on any of the four registers, then that SRAM is accessed and data is output with a fast address access and cycle time. When the comparator does not indicate a match the DRAM is accessed, loaded into the SRAM row register and available a given access time from row enable. In addition the hit pin indicates the miss so the processor is aware of the longer wait for data. Figure 6.53(a) shows the page mode read access for a hit and Figure 6.53(b) shows that for a miss.

Read accesses can be in either page mode or static column mode, as determined by the CAL pin. Since burst reads occur from the SRAM cache, the DRAM can be precharged at the same time effectively hiding the precharge time delay inherent in accessing a new row of a DRAM.

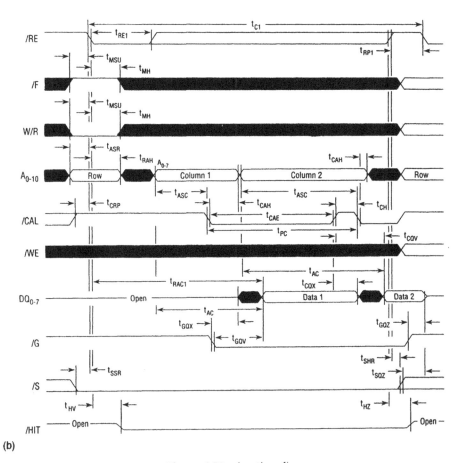

(b)

Figure 6.53   *(continued)*

The chip has an on-chip refresh counter and a dedicated refresh control pin which allows the DRAM array to be refreshed during cache read operations.

Memory writes are input through the data latch into the DRAM array. During a write hit, the on-chip address comparator activates a parallel write path to the SRAM cache which maintains coherency between the SRAM and the DRAM. Otherwise the data in the SRAM cache is not affected during write cycles.

The basic operating modes of the EDRAM are read hit, read miss, write hit, write miss, internal refresh, and low power standby.

The pinout of the EDRAM is significantly different from that of the standard DRAM. It more closely resembles that of the JEDEC SDRAM in adding additional power and ground pins to reduce ringing at high speeds. It also uses the 44-pin TSOP package like the SDRAM rather than the 32-pin TSOP used by the standard DRAM. There is also a 90° rotation of the address and I/O pins. The three packages are compared in Figure 6.54(a), (b) and (c).

The differences in pinout mean that a board designed for the standard asynchronous DRAM could not easily be used interchangeably for the EDRAM.

6.21.3   Enhanced Synchronous DRAM I

The enhanced synchronous DRAM is functionally a superset of the JEDEC Standard SDRAM I. This means that the ESDRAM can be used in a system as an SDRAM without the controller being aware that the superset exists. The ESDRAM is exactly

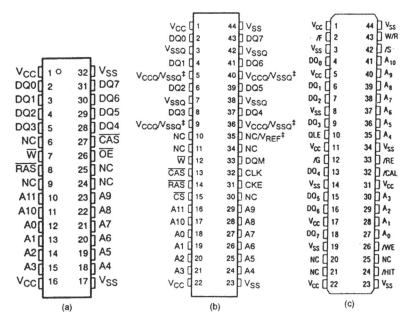

Figure 6.54 Packages of 512K×8 (a) DRAM, (b) SDRAM, (c) EDRAM (source: Texas Instruments, Ramtron)

pin-compatible with the SDRAM. The option in functionality is indicated in the mode register with the default being the standard SDRAM.

There are several modifications in the ESDRAM that differentiate it from the standard SDRAM. The first is that the RAS access time is faster than the typical SDRAM in the same technology. This reduces the latency when it is necessary to open a new page of the DRAM.

A second modification is that an SRAM cache is integrated onto the chip, as it is with the earlier EDRAM discussed in the previous section. This integrated cache consists of a row register for each DRAM bank of sense amplifiers as illustrated in Figure 6.55, which shows the block diagram for a four-bank ESDRAM part.

An on-chip comparitor, not available in the standard SDRAM, is used during a write-cycle to compare the last row read address with the current write address. The last row read address is the address of the data that is currently in the ESDRAM cache when the write is signalled.

Operationally, when a row in one of the DRAM arrays is opened, the row data is latched into the sense amplifiers, but is not transferred into the cache. The row could be closed again and refresh of another row done, or a different row opened without disturbing the data that is already in the cache SRAM.

If a read command occurs, then the data that is in the sense amplifiers is loaded into the cache. This permits the bank to be precharged on the next clock cycle without disturbing the data being read. For a standard SDRAM, the precharge cannot occur until after a number of cycles equal to the burst length. The ESDRAM recovery from a read page miss is, therefore, at least a burst length earlier than for the SDRAM. This ability to precharge earlier in the burst read cycle means that the gap between read bursts cen be reduced. This concept is illustrated in the read timing diagram in Figure 6.56.

A write is made directly to the bank of sense amplifiers, as illustrated in the block diagram. If a write command occurs after a new row is opened, the mode register is used to determine if the data in the sense amplifiers is loaded into the cache.

The mode register configuration for an ESDRAM determines whether the ESDRAM is in write transfer mode or not. If the mode register indicates that the ESDRAM is in "write transfer mode", then the write command causes the sense amplifier data to be loaded into the cache. The mode register configuration for an ESDRAM is shown in Figure 6.57.

If the ESDRAM is not in write transfer mode (in "no write transfer mode"), then the write command does not load the data in the sense amplifiers into the cache. This permits data from one row of the ESDRAM to continue to be available in the cache, while the data to a different row is updated. This means that write precharge and row access time can be hidden following a write command.

During a write cycle if the write page is already in the cache (a write hit), as determined by the on-chip comparitor, the ESDRAM will automatically update the cache when the data is written into the DRAM array, regardless of the state of the mode register.

The row data from the DRAM sense amplifiers is transferred into the cache only on the first read or write command after a bank activate command. Any subsequent read or write commands to the same row do not load the cache.

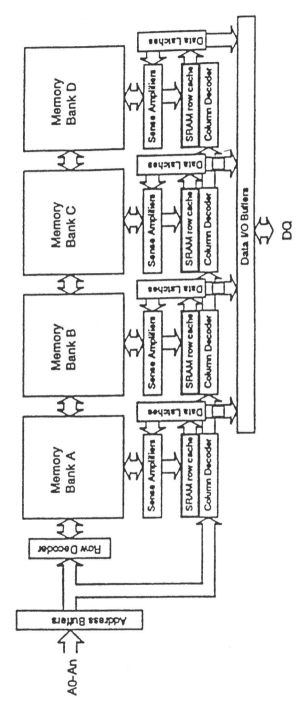

Figure 6.55   Block diagram of a four bank ESDRAM (source: Enhanced Memory Systems)

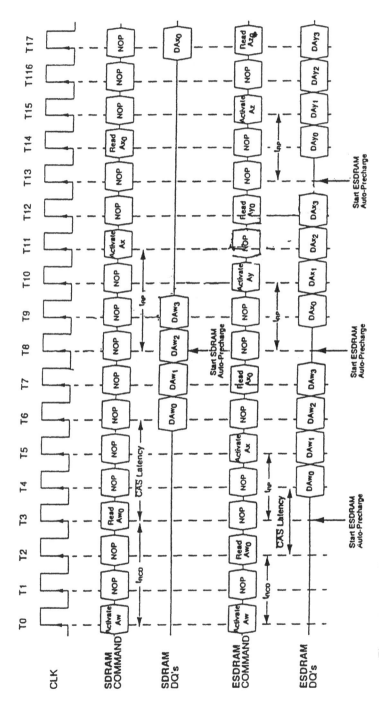

Figure 6.56 read timing diagram comparing the ESDRAM and the SDRAM (source: Enhanced Memory Systems)

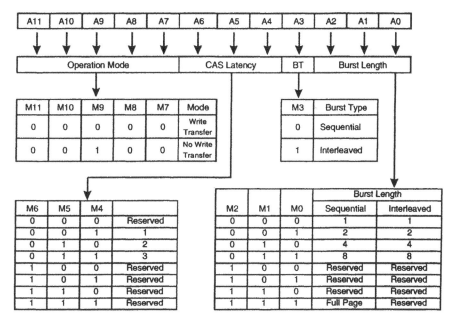

Figure 6.57   Mode register set for the ESDRAM ESDRAM 16M Datasheet [31] (source: Enhanced Memory Systems)

6.21.4   Enhanced Synchronous DRAM II (DDR)

The ESDRAM is also available as an SDRAM II compatible superset. Some of the operations of the ESDRAM I and II are compared in the timing diagrams in Figure 6.58, which show the read-autoprecharge operation for the ESDRAM and the DDR ESDRAM. In both cases, the precharge begins on the first clock cycle after the read command has been given.

In the "no write transfer" mode of operation the ESDRAM writes only to the DRAM array and not to the SRAM register in the case of a write page miss. This is the case when the page that is addressed is not the page loaded in the register. If the page addressed is the page loaded in the SRAM register, a write page hit, then both the DRAM and the SRAM cache are written.

Figure 6.59 shows the "no write transfer mode" operation for both the ESDRAM and the DDR ESDRAM. This mode assumes same bank operations, since fast alternate bank operations can be achieved with standard SDRAM.

At point "1" in the figure, Row x is activated and appears on the sense amplifiers. At point "2", a read is given and the data from the sense amplifiers (Row x) is transferred to the SRAM cache and after the two-cycle latency is read out in a four-bit burst of data. Meanwhile the array is precharged and row y is activated. The write command at point "3" is a "no write transfer mode" operation and the page that is being written (page y) is not the page that is in the SRAM cache (page x), so the ESDRAM writes only to the sense amplifiers, not to the SRAM. At point "4", a read command will access the data from row x that is in the SRAM cache.

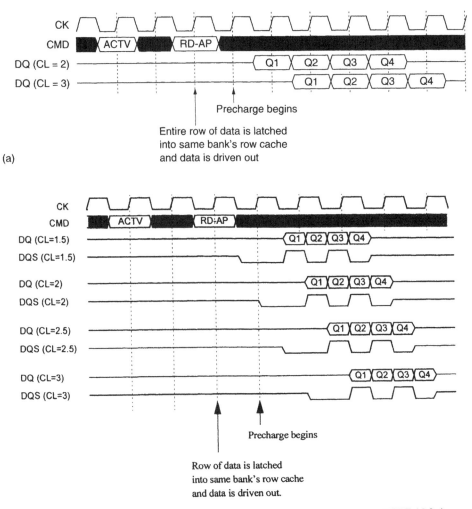

Figure 6.58   Read-autoprecharge for (a) SDR ESDRAM and (b) DDR ESDRAM (source: Enhanced Memory Systems Inc.)

6.21.5   The CDRAM

The cache DRAM, developed by Mitsubishi, is a synchronous, i.e. clocked, DRAM with an on-chip cache in the form of a separate SRAM array integrated on a chip with a DRAM array. Unlike the EDRAM, The SRAM and DRAM operations are separately controllable externally.

The CDRAM exploits the wider internal bus on chip to move data between the DRAM and the SRAM. This makes the CDRAM fast enough to be used as both main memory and second level cache in a low end unified system architecture.

For the 16M CDRAM, the DRAM is 1M×16 and the SRAM is 1K×16. The part has a programmable burst length of 4 or 8, and either a sequential or interleaved burst

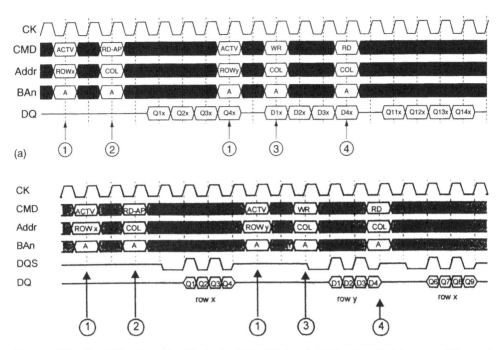

Figure 6.59   No Write Transfer Mode for ESDRAM and DDR ESDRAM (source: Enhanced Memory Systems Inc.)

sequence with no latency between bursts. It also has auto refresh and self refresh capability.

A block diagram of the 16M CDRAM is shown in Figure 6.60. The separate address inputs to the DRAM and SRAM are shown along with the read and write buffers.

Since the DRAM and the SRAM can be addressed independently, this permits the cache to be externally configured into either a direct mapped or set associate structure. It also makes it easy to add increments to the cache as DRAM is added to the system. The I/O structure is shared. Since the CDRAM is externally configurable, each added CDRAM increases the size of the system cache incrementally.

The access and cycle time options of the SRAM and DRAM on the 16M CDRAM are shown in Table 6.5.

A block data transfer between the DRAM and the data transfer buffers is performed in one instruction cycle. The four data transfer buffers – RB1, RB2, WB1, and WB2 – are shown in Figure 6.61 which is an operational schematic block diagram of the CDRAM.

The CDRAM has the advantage that a large block of data can be transferred from the DRAM to the SRAM in a single DRAM cycle. For the 16M CDRAM, the block size is 8×16. This counters a drawback found with external L2 caches, that a large block size increases the miss penalty due to the delay in transferring a large block from the DRAM to the SRAM over external wiring. For this reason most external caches have relatively small block sizes.

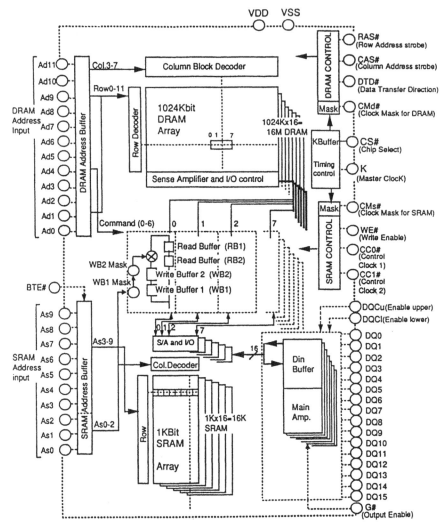

Figure 6.60   Block diagram of the 16M CDRAM (source: Mitsubishi [18])

Table 6.5   Access and cycle time options of the SRAM and DRAM on the 16M CDRAM

Selection	SRAM		DRAM	
	t_{AC}	t_{CYC}	t_{AC}	t_{CYC}
−10	10	10	60	120
−12	12	12	60	120
−15	15	15	70	130

Source: Mitsubishi (M5M4V16169TP Target Spec. Rev 4).

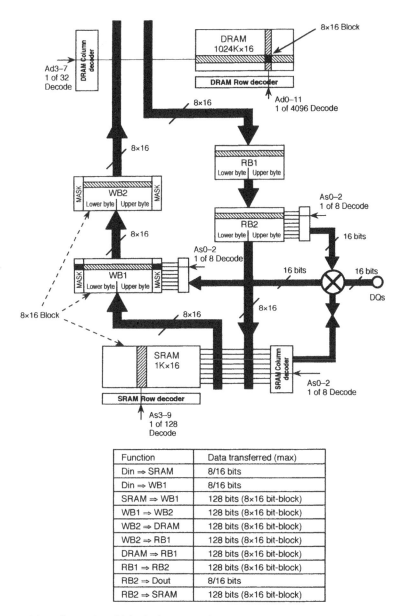

Figure 6.61 Operational block diagram of the 16M CDRAM (source: Mitsubishi [18])

However, it is a fact of cache architecture that a small cache with a large block size, such as on the CDRAM, will have a comparable miss ratio to a large cache with a small block size. This means that the miss ratio can be maintained by going to a small CDRAM cache without incurring a larger miss penalty.

Caches are effective because of both spatial and temporal locality of many programs. Because L1 caches on the microprocessor chip are relatively small, they exploit primarily temporal locality. That is, if the processor returns repeatedly to the same location in memory in a short period of time as in a program loop then that one location can be stored in a very small cache and be readily available.

The larger CDRAM cache used in conjunction with a small L1 cache on the processor primarily exploits spatial locality which is the tendency for the processor to ask for a next data location very close to the previous data location. This is because a larger block of data can be transferred into the CDRAM cache on each request to the main memory. The two functions are therefore complementary.

The cache tag of the CDRAM needs to be supplied separately from the CDRAM chip. This is a trade-off. The CDRAM has more flexibility in cache configuration because of the off-chip control; however, it would be simpler and faster if integrated onto the CDRAM. It could also be integrated into the processor; however, processor chips are themselves commodity products and DRAM specific processor chips are rare. It is more likely that the tag for the CDRAM cache will need to be supplied with a small logic chip.

Coherency between the cache (SRAM) and the main memory (DRAM) is maintained using fast copy-back. When a cache miss occurs the old data is moved into a temporary buffer on chip while a new block is accessed from the DRAM into the cache. This access takes 70 ns. If the old block needs updating, it is moved to the DRAM memory where it is updated and returned to the cache while the processor is accessing the cache. This method hides the latency of write-back cycles unless there are back-to-back misses.

A 70-pin package is required with pinout as shown in Figure 6.62. To reduce ground bounce effects during high speed transfers there is one power/ground pin for every two I/O pins.

The separate address inputs for the DRAM and the SRAM permit fast SRAM access; however, they do require additional pins on the package over those required for a pure SDRAM.

While the chip can function in purely synchronous mode, it also has selectable output operation to permit transparent, latched or registered outputs. It also has asynchronous output enable for bus control.

The chip is synchronous under the control of the system clock. It operates as a state machine with a truth table which is shown in Figure 6.63.

Since all the possible operations between the DRAM, the SRAM, and the on-chip read and write buffers are externally controllable, this part has a great deal of flexibility in selecting operating modes.

Illustrations for a few of the modes are shown in Figure 6.64 (see pages 209 and 210).

A buffer write transfer and SRAM write is illustrated in Figure 6.64(a). In this mode the data is first written to the SRAM from the I/O pins using the SRAM address inputs As0–As9. The buffer write transfer cycle then clears all transfer mask bits in the WB1 mask which permits all data to be transferred in a successive DRAM write transfer cycle.

A buffer read transfer and SRAM read cycle is illustrated in Figure 6.64(b). In this mode the data is transferred from the read buffer (RB2) to the SRAM and simultaneously data is read from the RB to the I/O pins. SRAM addresses As3–As9 select

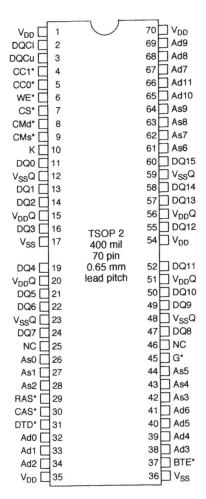

Figure 6.62   Pinout of 16M CDRAM (source: Mitsubishi [18])

the SRAM row to which the 8x16 bit block is to be written. Addresses As0–As2 decode the 16-bit word to be read.

A DRAM write transfer 1 and read is shown in Figure 6.64(c). In this mode an 8×16-bit block of data is transferred from WB1 through the WB2 buffer to the DRAM block specified by the DRAM addresses Ad3–Ad7. Addresses Ad8–Ad11 are set low. The mask present in WB1 is also transferred to WB2 and controls the data written to the DRAM. The block to which the data is written in the DRAM is simultaneously transferred to the read buffers.

The CDRAM could be useful in some mid- to low end computer systems by permitting a unified architecture with only processor and DRAM, eliminating the cost and space of the separate L2 cache.

In other low end systems with graphics but normally without L2 cache, it might be able to give some performance improvement to the L1 cache while primarily serving

CODE	CS#	Prev CMs#	CC0#	CC1#	DOC(µ0)	WE#	AsS (As0-9)	Prev CMd#	RAS#	CAS#	DTD#	Ad0-9 (DRAM addr)	Ad2:/Ad7:Ad00	WB1	WB2	WB1 Mask	WB2 Mask	Read Buffer	Din	Dout	Function
NOP	H	H	X	X	X	X	X	H	X	X	X	X		-	-	-	-	-	-	HI-Z	No OPeration
SPD	X	L	X	H	X	X	X	X	X	X	X	X		-	-	-	-	-	-	Suspend	SRAM Power Down & Data retention (No operation)
DES	L	H	H	H	H	X	X	X	X	X	X	X		-	-	-	-	-	-	HI-Z	Deselect SRAM (No operation)
SR	L	H	H	L	H	H	As0-9	X	X	X	X	X		-	-	-	-	-	-	Valid	SRAM Read (SRAM->DO)
SW	L	H	H	L	H	L	As0-9	X	X	X	X	X		-	-	-	-	-	Valid	HI-Z	SRAM Write (DIN->SRAM)
BRT	L	H	L	H	L	H	As3-9 (2)	X	X	X	X	X		-	-	-	-	Use	-	HI-Z	Buffer Read Xfer (RB2->SRAM)
BWT	L	H	L	H	L	L	As3-9 (2)	X	X	X	X	X		Load	-	Clear Mask	-	-	-	HI-Z	Buffer Write Xfer (SRAM->WB1)
BRTR	L	H	L	H	H	H	As0-9	X	X	X	X	X		-	-	-	-	Use	-	Valid	Buffer Read Xfer & Read (RB2->SRAM->DO)
BWTW	L	H	L	H	H	L	As0-9	X	X	X	X	X		Load	-	Clear Mask	-	-	Valid	HI-Z	Buffer Write Xfer & Write (DIN->SRAM->WB1)
BR	L	H	L	L	H	H	As0-2 (2)	X	X	X	X	X		-	-	-	-	Use	-	Valid	Buffer Read (RB2->DO)
BW	L	H	L	L	H	L	As0-2 (2)	X	X	X	X	X		Load	-	Clear (4) or 2 bit	-	-	Valid	HI-Z	Buffer Write (DIN->WB1)
DPD	X	X	X	X	X	X	X	L	X(1)	X(1)	X(1)	X		-	-	-	-	-	-	-	DRAM Power Down (No operation)
DNOP	L	X	X	X	X	X	X	H	H	H	X	X		-	-	-	-	-	-	-	DRAM No OPeration (No operation)
DRT	L	X	X	X	X	X	X	H	L	L	H	Ad0-7 (2) (Col.Block)	0:0:0	-	-	-	-	Load	-	-	DRAM Read Xfer (DRAM->RB1->RB2)
DWT1	L	X	X	X	X	X	X	H	L	L	L	Ad3-7 (2) (Col.Block)	0:0:0	Use	Load/Use	Use(6)/Load	-	-	-	-	DRAM Write Xfer1 (WB1->WB2->DRAM) (3)
DWT1R	L	X	X	X	X	X	X	H	L	L	L	Ad3-7 (2) (Col.Block)	0:0:1	Use	Load/Use	Use(6)/Load	-	Load	-	-	DRAM Write Xfer1 & Read (WB1->WB2 ->DRAM->RB1->RB2) (3)
DWT2	L	X	X	X	X	X	X	H	L	L	L	Ad3-7 (3) (Col.Block)	0:1:0	-	Use	-	Use	-	-	-	DRAM Write Xfer2 (WB2->DRAM)
DWT2R	L	X	X	X	X	X	X	H	L	L	L	Ad3-7 (2) (Col.Block)	0:1:1	-	Use	-	Use	Load	-	-	DRAM Write Xfer & Read (WB2->DRAM ->RB1->RB2)
ACT	L	X	X	X	X	X	X	L	H	H	H	Ad0-9 (Row Add.)		-	-	-	-	-	-	-	DRAM Activate (Page Call)
PCG	L	X	X	X	X	X	X	L	L	H	H	X		-	-	-	-	-	-	-	DRAM Precharge
ARF	L	X	X	X	X	X	X	H(7)	L	L	H	X		-	-	-	-	-	-	-	Auto Refresh
SRF	L	X	X	X	X	X	X	H(8)	L	L	H	X		-	-	-	-	-	-	-	Self Refresh Entry
SCR	L	X	X	X	X	X	X	H	L	L	L	Command		-	-	-	-	-	-	-	Set Command Register

Figure 6.63   Truth table for 16M CDRAM (source: Mitsubishi [18])

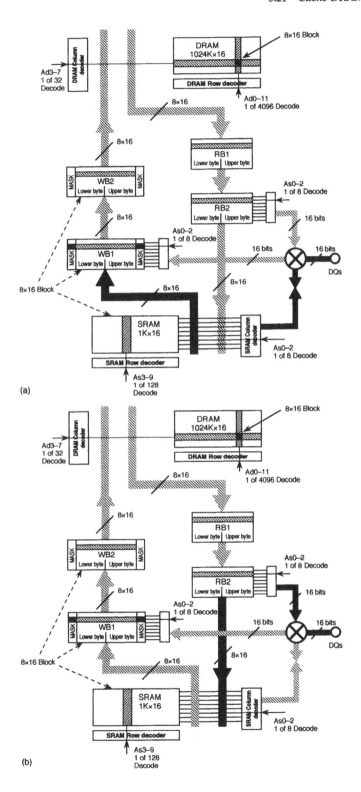

(a)

(b)

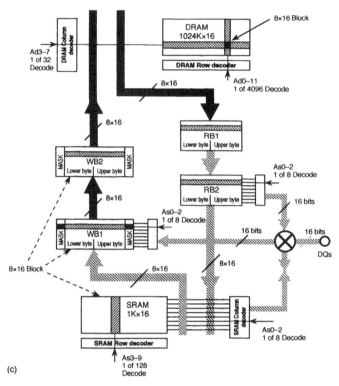

Figure 6.64   Operating modes for the 16M CDRAM: (a) buffer write transfer and SRAM write; (b) buffer read transfer and SRAM read; (c) DRAM write transfer 1 and read (source: Mitsubishi [18])

to unify the main memory and graphics. Systems of this type might include price-sensitive consumer-oriented systems such as games, PDAs, subnotebook PCs, set top decoder boxes, and multimedia PCs.

Other cache DRAMs have also been shown. A 125 MHz 256Mb synchronous cache DRAM from Oki has been described which integrates both the cache SRAM and the tag onto the DRAM chip [19].

The DRAM array is divided into 32 banks of 1MB each. A 32KB cache SRAM is associated with each DRAM bank. The cache block size is 2KB. The RAS latency for a cache hit is 23 ns and for a bank hit is 31 ns. The CAS latency is 15 ns.

Bank interleaving is used to improve datarate. One bank controller is associated with each bank to control the memory operations in that bank such as cache hit or miss judging processes. The controllers are distributed through the array with each located near its associated bank. Each bank has an associated tag which holds the row address of the preceding request addressed to a location in the associated bank.

6.21.6   Virtual Channel SDRAM

The Virtual Channel SDRAM is another synchronous DRAM with cache SRAM added to improve the access time. This part, which was developed by NEC, is equivalent to a 16-way set-associative cache. There are 16 SRAM registers, called "channels" each of which is the size of $\frac{1}{4}$ of a DRAM sense amplifier bank. The DRAM array has a two-bank architecture. Each row of a DRAM bank is split into four segments and each of these segments can be one transfer operation to or from one of the SRAM registers.

A block diagram of the VC-SDRAM is shown in Figure 6.65.

This part is a superset of the standard SDRAM. Several special operations are available. These include: uploading data from the memory core to the channel, uploading two segments of data from the memory core to two channels with one command, uploading data to the channel and reading in one command, and downloading data from the channel to the DRAM core. it is not possible to write directly to the DRAM core bypassing the SRAM channels.

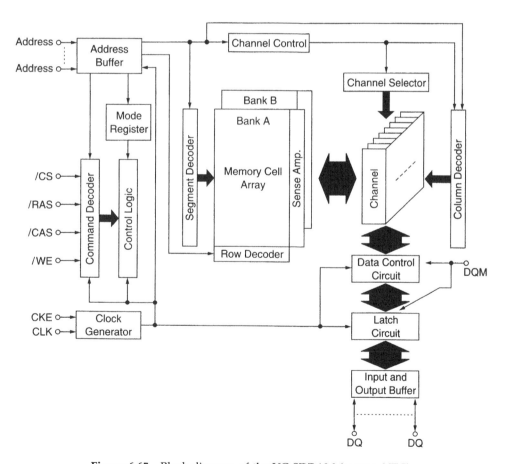

Figure 6.65   Block diagram of the VC-SDRAM (source: NEC)

A conceptual schematic illustrating this 16-way set associative architecture is shown in Figure 6.66.

Operations such as prefetch and restore are done in the background between the DRAM array and the channels. Restore is the operation that moves data from the SRAM channel to one of the sense amplifier segments in the DRAM array. Read and write operations are done between the input/output bus and the SRAM channels.

This part might be particularly useful in a multitasking environment, where several memory masters could have effective access to the DRAM core through their own SRAM channels.

6.22   Packet Protocol Synchronous DRAMs

The packet protocol synchronous DRAMs are synchronous DRAMs where all of the information required for an operation is configured in a digital packet which is transferred over a high speed bus to all the DRAMS on a module. All resulting operations on the DRAM are controlled within the DRAM itself. During a read operation, data from the DRAM is issued in another packet over a high speed bus from the DRAM back to the controller. For a write, the controller issues the data packet.

The packet protocol parts, by moving all DRAM control onto the chip itself, result in a very fast transfer of commands and data between the DRAM and the controller. An additional latency is, however, involved in the use of the packet protocol.

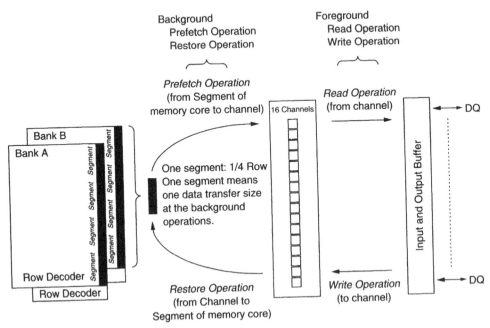

Figure 6.66   Conceptual schematic showing set-associative nature of 64M VC-SDRAM architecture (source: NEC)

6.22.1   Packet Protocol Synchronous DRAMs from Rambus, Inc.

Rambus, Inc. has developed three generations of packet protocol synchronous DRAMs. The first generation part was used primarily in a gaming application. The second generation part, called the Concurrent RDRAM, was targeted primarily at graphics applications. The third generation part, called the Direct Rambus DRAM was jointly developed with Intel for use in PCs.

First generation Rambus DRAM

The first generation Rambus DRAM was primarily targeted at graphics applications and is described in detail in Chapter 7.

A block diagram of the first generation 18M Rambus DRAM is shown in Figure 6.67.

The underlying architecture of the 18M Rambus DRAM shows its similarity to the 16M SDRAM. The uppermost part of the diagram, referred to as the "application

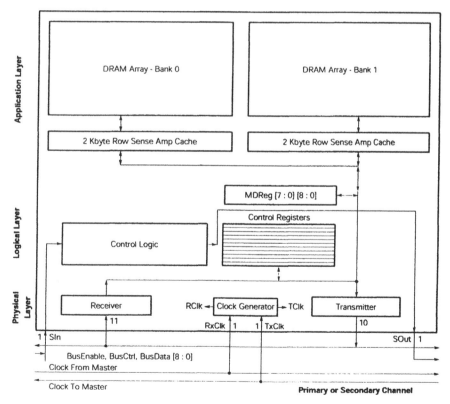

Figure 6.67   Block diagram of the first generation 18M Rambus DRAM (source: NEC 18M Rambus DRAM Datasheet)

layer", shows two DRAM arrays with two banks containing 4K bytes of sense amplifiers, which are referred to as "sense amplifier caches". Below the sense amplifiers are shown the control logic and registers that allow the underlying asynchronous DRAM to operate synchronously using the packetized command and data structure. This is referred to as the logical layer. The packet protocol interface to the bus, called the "physical layer", is shown below that. This layer consists of receivers, transmitters, and clock generator with clock-from and clock-to the master controller, as well as the serial "daisy chain" link.

Concurrent Rambus DRAM

The second generation 64M Rambus DRAM is called the Concurrent Rambus DRAM. Like other 64M synchronous DRAMs, the part has four banks which permit concurrent operations in different banks. A block diagram of the 64M concurrent Rambus DRAM is shown in Figure 6.68.

This 3.3 V part is specified with a 600 MB/sec peak transfer rate for each RAM. The effective bandwidth for random 32 byte transfers from one RDRAM is 480 MB/sec. It continues to have the combined command and databus used in the first generation Rambus DRAM.

Direct Rambus DRAM

The Direct Rambus DRAM is the third generation Rambus part. It was cooperatively developed by Rambus, Inc. and Intel Corporation and has been selected by Intel for use with various Intel PCs expected to be available in 1999 or thereafter.

This packet protocol 64M/72M-bit part is specified to run at both 600 and 800 MHz on the 16-bit databus. The resulting bandwidth is 1.2 GB/sec and 1.6 GB/sec. This is an equivalent bandwidth to a DDR SDRAM running at 300 MHz and 400 MHz datarate on a 32-bit bus. The interface continues to be the RSL (Rambus Signalling Level). The part is packaged in a ball grid array (BGA) type of package.

A block diagram is shown in Figure 6.69.

The Direct Rambus DRAM uses a 16-bit databus comprised of two 8-bit databuses over which a "dualoct" (16 bits) of data is transferred to and from the RAM array. It has separate command and databuses and separate row and column control buses. It is divided into 16 banks which are accessed via shared sense amplifiers giving it an eight independent bank operation. Four transactions can occur simultaneously.

A 72-bit wide internal bus, which runs at $\frac{1}{8}$th the frequency of the data, demultiplexes to the two internal 8-bit databuses which are fed out at full speed to the external databus. This permitis 100 MHz internal operation to appear on the external bus at 800 Mbs/pin. Two write buffers are included for temporary storage of write data. In addition to storing a 16-bit data word, the write buffer can also store the bank column address associated with the stored word.

The control registers, which are used to select the operating mode, are shown in the upper part of the block diagram along with the pins contolling them, CMD, SCK, SI01 and SI02. The clock-to-master pins (CTM and CTMN) generate both the internal

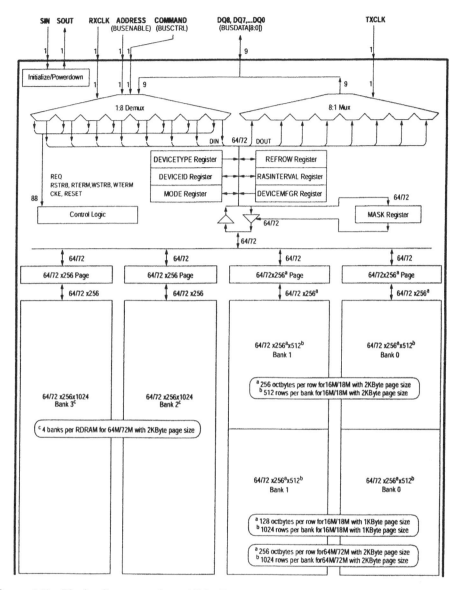

Figure 6.68 Block diagram of a 64M Concurrent Rambus DRAM (source: Rambus Concurrent DRAM Datasheet [31])

receive clock used for write data and the internal transmit clock used for read data. The three ROW pins are used for the bank activate and the five COL pins are used for the read and write commands.

The Direct Rambus DRAM contains seventeen 512-byte sense amplifiers, each holding one half of one row of a single bank of the DRAM. Since the sense amplifiers are shared between two banks, adjacent banks may not be simultaneously accessed.

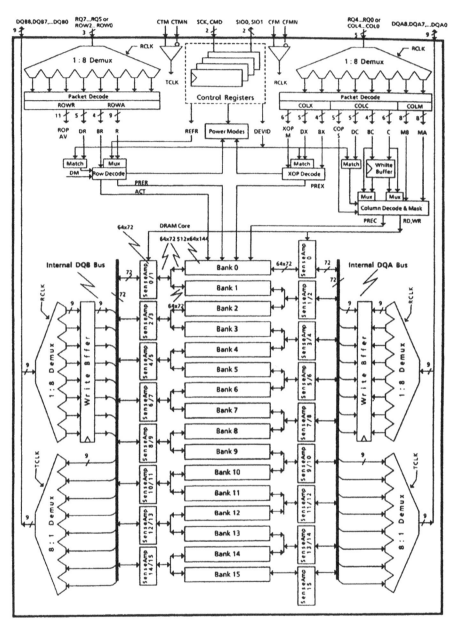

Figure 6.69   Block diagram of the 64M/72M Direct Rambus DRAM (source: Toshiba direct RDRAM Datasheet [30])

The Direct Rambus DRAM architecture includes 16-RAM arrays which are accessed by shared sense amplifiers. These arrays are organized 512×64 with a 72-bit internal bus on which eight 8-bit words can be prefetched at one time on two databuses, the DQA0-8 and DQB0-8.

A single page has 64 words available on two open sense amplifiers banks at one time. The two internal 72-bit data paths can be used for two 8-bit data inputs, which are clocked in as a 16-bit word by the receive clocks, and two 8-bit outputs, which are clocked out as a 16-bit word by the transmit clocks.

The write command causes a 16-bit word of data to be loaded into the write buffer. The write buffer has been added to improve the efficiency of a write followed by a read to a different address. The write buffer acts as a write cache. Data stored there can be written to the array on a subsequent cycle. It acts, therefore, as a copy-back cache. Writing a word stored in the write buffer into the array is referred to as "retiring" the write buffer.

Commands are on a separate bus from the data. The row and column controls are internally separate. Both row and column commands are input as 8-bit words which are demultiplexed and decoded. The resultant command is then clocked into the DRAM array.

Row commands are demultiplexed into two operations: Row Activate (ROWA) and Row Operate (ROWR). A row activate command from a ROWA pocket opens one of the 512 rows and loads it into its two associated 256-bit sense amplifier banks for each of the DQA and DQB databuses. A precharge (PRER) command from a ROWR (row retire) packet closes the sense amplifiers associated with a particular array so a different row can be activated or another bank can be activated.

A read (RD) command causes one of the dualocts from the open banks of sense amplifiers to be transmitted over the 8-bit DQA and DQB buses. A Write (WR) command causes a 16-bit (dualoct) received from the external bus to be written into the write buffer.

There are also Read with Auto-precharge (RDA) and Write with Auto-precharge (WDA) commands. These commands cause the sense amplifier bank to be closed after the read or write operation is complete. In the case of the write operation, the write buffer needs to be retired before the row is closed, to avoid accidently writing the buffer to the wrong row.

The column pins are demultiplexed into either a 23-bit column operation packet (COLC), a 17-bit column mask (COLM) packet or a 17-bit extended operation (COLX) packet.

The commands and data are transferred between the Rambus DRAM and the RIMM controller in packets which are clocked on the rising and falling edge of the system clock. The clock maximum speed is 400 MHz, so that the packet information is transferred at an 800 Mps/pin rate. All the information that is required to identify the proper DRAM on the module and give it the desired command and correct row and column addresses are included in such packets.

Figure 6.70 illustrates examples of some types of packets available, along with a timing diagram using these packet types. The set-ups shown are of: a row activate (ROWA) packet; a write command (COLC) packet; a precharge command (ROWR) packet; another column packet used for masking (COLM); and a column packet (COLX) which can be used for external precharge (PREX) or other housekeeping types of commands.

The data packet is transferred on a separate bus. All packets on the row and colomn pins use the trailing edge of the packet as a reference point, whereas data packets use the leading edge of the packet as a reference point.

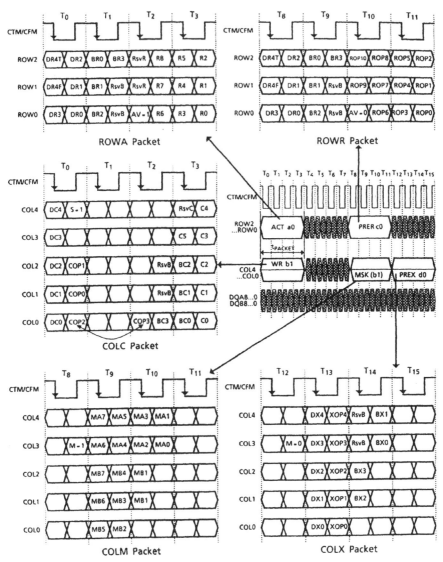

Figure 6.70 Packet configuration for a Direct Rambus DRAM (source: Toshiba Direct RDRAM Datasheet [30])

The normal underlying asynchronous DRAM timing constraints are obeyed in the Direct Rambus DRAM as shown in Figure 6.71, which illustrates the timings associated with two back-to-back read operations.

The ACT command packet (row activate) is followed after the minimum t_{RCD} timing by a column RD command packet (column read) which is followed after a time TCC by another column read packet. After a time t_{CAC} (CAS latency) following the first read command, the data packet for the first read command is received and

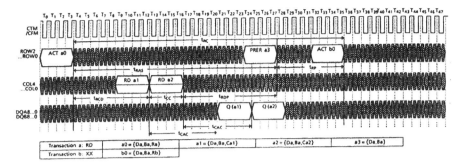

Figure 6.71 Timing diagrams for two back-to-back read operations (source: Toshiba Direct RDRAM Datasheet [30])

after a time t_{CAC} after the second read command, the associated data packet is received. A time t_{RDP} after the read command, which gives time for the burst of data to be through the internal bus, the precharge command is issued, and then after the precharge is completed (TRP), another row of the bank can be opened using another activate command.

A write transaction is a bit more complex than in the SDRAM, since the addition of the write buffer means that the write buffer must be written then retired before the bank of sense amplifiers is closed. An example of two back-to-back writes is shown in Figure 6.72.

The row activate packet is followed after a time t_{RCD} by the first write command packet. The data for this command begins to be written in a write latency time t_{CWD} after the write command packet. The write commands can be followed after a minimum time t_{CC} by a write buffer retire command or mask command.

A problem can occur if care is not taken to retire the write buffer before closing the sense amplifier banks and opening another bank. In this case, when the write buffer is closed, it will be retired to the wrong bank of the part, resulting in incorrect data in both the original row and the newly retired row.

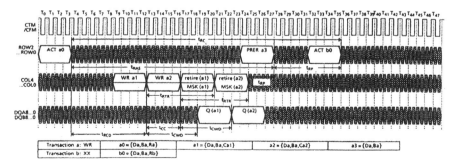

Figure 6.72 Two back-to-back writes for the Direct Rambus DRAM (source: Toshiba Direct RDRAM Datasheet [30])

Other command configurations can be found in the Direct Rambus Datasheet.

The Direct Rambus DRAM is delivered in a Ball Grid Array (BGA) package, also called a "chip scale" package. The top view of a pinout for the array is shown in Figure 6.73. The top view is taken by convention as looking through the package at the balls.

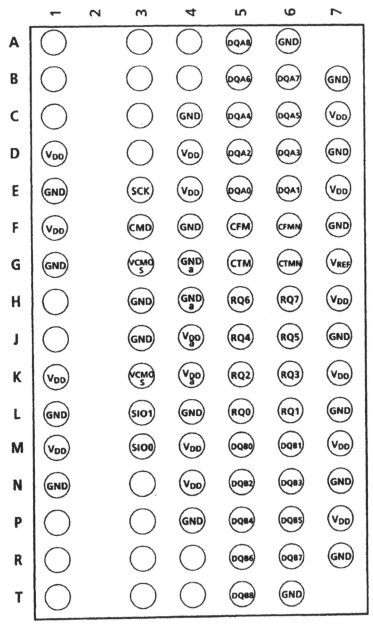

Figure 6.73   Pin diagram of the top view of a Direct Rambus DRAM in BGA package (source: Toshiba Direct DRAM Datasheet [30])

6.22.2   Synchronous Link DRAM (SLDRAM)

The Synchronous DRAM is an implementation of the Ramlink DRAM protocol which was passed by the IEEE P1596.4 Committee.

Ramlink is a high performance protocol based on event scheduling. The Ramlink logical protocol is shown in Figure 6.74. Requests and responses are issued in a non-blocking protocol with concurrent read and write.

The SLDRAM implementation is intended to support bursts of 500MHz and beyond. It consists of a specialized link controller, several SLDRAMs on one-way write and read buses and the interconnects between the devices.

The Synchronous Link DRAM, or SLDRAM, is a synchronous DRAM which has a packet protocol interface. It is a direct evolutionary step from the DDR SDRAM. It requires a dedicated controller which packages all the input commands and addresses in a single packet of information, this is then transmitted on a low swing, high speed bus to all SLDRAMs on a module. An identifier bit in the packet permits the addressed DRAM to identify itself and to begin processing the commands in the packet. The data travels on a separate bus to and from the controller. One advantage of the packet protocol interface is that it is compatible across multiple densities of DRAMs.

The high speed bus permits burst of data to be transferred at a targeted rate of up to 800 Mbs/pin. Like the DDR SDRAM, data is transferred on the rising and falling edge of the command clock. The data transfer speed possible on the 16-bit databus is 1.6Gbs/pin.

While a DLL (delay lock loop) is used on the SLDRAM, there is no PLL used. The part has 8 banks, each having 2k rows and 64 columns. A fixed burst length of 4 is specified for the command burst. The internal databus is 72 bits wide and demultiplexed down to an external 16-bit databus. A four-word prefetch buffer in the data path is used to hold the 72-bit word. The data burst can be 4 or 8.

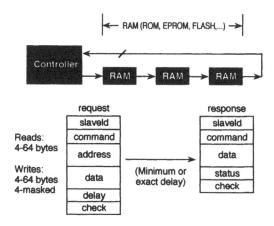

Figure 6.74   Ramlink logical protocol

The high speed bus requires a specialized controller which is called a SLAC (SLDRAM Controller). The command bus has a free running differential clock. Commands are transferred in packets. The databus has an intermittent differential clock which transfers data between the controller and the selected DRAM.

The command link uses a 10-bit command/address, and a flag which indicates the start of a command packet. The data link has 16 or 18 bits of data. In addition there is a CMOS daisy-chained link which is used for various housekeeping functions.

The SLDRAM module has one memory controller and up to eight loads. The loads can be either discrete SLDRAMs or buffered modules with many SLDRAM devices.

The SLDRAM devices can be packaged in either 0.5 mm pitch TSOP or 0.8 mm staggered pitch vertical surface mounted packages (VSMP). The SLDRAM pinout in VSMP is shown in Chapter 9.

The commands and data for the 64M SLDRAM are formulated in four 10-bit packets. The command packet structure is shown in Figure 6.75. The packet consists of a flag bit, ID bits which identify which SLDRAM on the module is addressed, command (CMDx) bits, Bank select bits (BNKx), Row address bits (ROWx), and Column address bits (COLx).

For a read or write for the SL-DRAM, the flag signal indicates the start of a valid request packet. The data clocks (DCLK) are similar to the data strobes used by the DDR SDRAMs and have a similar preamble and postamble.

The controller can program the read and write latency of each of the SLDRAMs in the module. This permits the contoller to compensate for varying performance of SLDRAMs from different manufacturers, and for varying positions of the SLDRAMs on the bus. The read latency can be adjusted in fine and coarse increments, while the write latency can be adjusted only in coarse increments.

The SLDRAM controller is also able to calibrate the VOH and VOL levels of the individual SLDRAMs. The Datalink daisy chain is used for this operation. The SLDRAMs drive the Datalink with high and low levels, which can be incremented and decremented until they match with the reference levels in the controller.

The read latency can also be synchronized for the different SLDRAMs by incrementing the internal read data vernier present in each SLDRAM architecture.

FLAG	CA9	CA8	CA7	CA6	CA5	CA4	CA3	CA2	CA1	CA0
1	ID8	ID7	ID6	ID5	ID4	ID3	ID2	ID1	ID0	CMD5
0	CMD4	CMD3	CMD2	CMD1	CMD0	BNK2	BNK1	BNK0	ROW9	ROW8
0	ROW7	ROW6	ROW5	ROW4	ROW3	ROW2	ROW1	ROW0	0	0
0	0	0	0	COL6	COL5	COL4	COL3	COL2	COL1	COL0

Figure 6.75   SLDRAM command packet structure (source: SLDRAM Consortium)

Bibliography

1. Wilson, R., JEDEC hustling to spec new SDRAM, *Electronic Engineering Times*, 23 March 1992, 1.
2. JEDEC JC42.3 press release on the synchronous DRAM, 2 June 1992, EIA JEDEC.
3. Powell, E. W., and Vogley, W.C., Synchronous DRAMs break the memory system bandwidth bottleneck, 1992 PC Design Conference paper, August 1992.
4. Prince, B., *Semiconductor Memories,* 2nd Edition, John Wiley and Sons, 1993.
5. Jones, F., A new era of fast dynamic RAMs, *IEEE Spectrum*, October 1992, 43.
6. Hart, C.A., Dynamic RAM as secondary cache, *IEEE Spectrum*, October 1992, 48.
7. Hart, C.A., *CDRAM in a Unified Memory Architecture,* Mitsubishi Applications Note, 1995.
8. *Ramtron Specialty Memory Products Data Book*, October 1994.
9. Takai, Y. *et al.*, 250 Mbytes/s synchronous DRAM using a 3-state-pipelined architecture, *IEEE Journal of Solid State Circuits*, Vol. 29, No. 4, April 1994.
10. Furuyama, T. *et al.*, A high random access data rate 4Mb DRAM with pipeline operation, *Symposium on VLSI Circuits*, 1990, 9.
11. Prince, B. *et al.*, "synchronous dynamic RAM, *IEEE Spectrum*, October 1992, 44.
12. *Toshiba Applications Specific DRAM Databook*, 1994, D-20.
13. Choi, Y. *et al.*, 16Mb synchronous DRAM with 125-Mbyte/s data rate, *IEEE Journal of Solid State Circuits*, Vol. 29, No. 4, April 1994.
14. Kodama, Y., A 150 Mhz 4-Bank 64M-bit SDRAM with address incrementing pipeline scheme, *Symposium on VLSI Circuits Digest of Technical Papers*, 1994.
15. Prince, B., Speeding up system memory, *IEEE Spectrum*, February 1994, 38.
16. Hoi-Jun Yoo *et al.*, A 150 MHz 8-Banks 256M synchronous DRAM with wave pipelining methods, *Proceedings of the IEEE ISSCC*, February 1995, 250.
17. *Texas Instruments MOS Memory Databook*, 1993.
18. *Mitsubishi M5M4V4169 Cache DRAM Datasheet.*
19 Tanoi, S. *et al.*, A 32-bank 256-Mb DRAM with cache and TAG, *IEEE Journal of Solid State Circuits*, Vol. 29, No. 11, November 1994, 1330.
20. *16M Synchronous DRAM Datasheet*, Mitsubishi, 1995.
21. *Standard for Semiconductor Memories*, JESD21C, Electronic Industries Association, JEDEC.
22. *Proposal for Synclink, Hyundai*, 1995.
23. *Toshiba 64M SDRAM I Datasheet*, <www.toshiba.com>.
24. *Mitsubishi 64M SDRAM I Datasheet*, <www. mitsubishi.com>.
25. *Mitsubishi 128M SDRAM (SDR) Datasheet*, <www.mitsubishi.com>.
26. Toshiba 128M DDR SDRAM Datasheet <www.toshiba.co>
27. Gillingham, P., SLDRAM architectural and functional overview, *SLDRAM web page*, <www.sldram.com>, August 29, 1997.
28. *SLDRAM 64M Datasheet, SLDRAM Consortium,* <www.sldram.com>.
29. *16M ESDRAM Datasheet, Enhanced Memories Inc.,* <www.esdram.com>.
30. *Direct RDRAM Datasheet, Toshiba,* <www.toshiba.com>.
31. *Concurrent RDRAM Datasheet, Rambus,* <www.rambus.com>.
32. *18M Rambus DRAM Datasheet, NEC,* <www.nec.com>.
33. *64M Virtual Channel DRAM Datasheet, NEC,* <www.nec.com>.

7 Graphics DRAMs

7.1 Overview of DRAMs for Graphics Subsystems

Graphics in television systems and graphics subsystems in computers use a sufficient amount of memory to require the higher density of DRAMs. These subsystems also have special requirements for higher bandwidth than is available on the basic asynchronous DRAM.

A basic frame buffer in either the video or graphics application is required to provide at a minimum a continuous serial stream of data to refresh the display screen. A subsystem with some manipulation of bits requires, in addition, a random port for fast interface with the processor or graphics controller to provide the required data manipulations. Either these specific graphics requirements need to be met on the memory chip or a very high bandwidth memory must be available to support a graphics controller and a parallel-to-serial device which provide the required functions. Various combinations of these two approaches have been tried. These approaches are outlined in this chapter.

The market volumes of the systems involved are historically high enough to have generated a number of applications-specific DRAMs to serve the special requirements of the graphics subsystems. These included through the 1980s a variety of simple serial frame memories used in television applications and the more standard dual ported video DRAMs which have one serial and one random port which have been used in computer applications requiring more graphics manipulations.

In the 1990s the datarate of the single port DRAM has been increased significantly with the introduction of the EDO (Hyperpage) mode and synchronous DRAMs, with very wide interfaces. This has led to increased use of fast, wide single port DRAMs in graphics subsystems in PCs coupled with the use of standard graphics controllers which provide the multi-port interface to the processor and the RAMDAC. Many new single port DRAM variants are also being developed such as the synchronous graphics DRAMs (SGRAM), Double Data Rate (DDR) SGRAM, the Multibank DRAM and the Dual Port SDRAM.

Multi-port VRAMs continue to be used in high end applications and appearing on the horizon are chips which gain performance and save space by integrating larger parts of the graphics subsystem, either the logic and DRAM separately or the entire subsystem, into the DRAM. These integrated chips are being initially developed for portable applications where the lower power consumption of the integrated DRAM with logic circuit helps extend battery life.

Meanwhile for low end PC applications there is a trend toward unified memory which returns the frame buffer function to the main memory of the computer. Unified memory is made possible by advances both in processor and DRAM bandwidth and in software development.

This chapter considers first simple serial video memories for television applications, then the high bandwidth single port graphics memories, the multi-port memories, and finally the new phenomenon of the integrated graphics memory.

7.2   Frame Memories for Television Applications

7.2.1   *Simple Serial Field Memory for Temporary Frame Storage*

In a basic video subsystem such as is used in television sets to reduce visible lines or visible flicker on the screen, only a simple serial access storage device with four-bit input and output is required to store a frame and recycle it to the screen. A random access port is not required since there is no graphics manipulation involved.

Such devices are called field memories and are frequently made from DRAMs configured with serial input and output ports instead of a single parallel port. The DRAM core array is not changed. Read and write frequencies range from 33 to 50 MHz for a line of serial data.

An example is a 256K×4 field memory from Texas Instruments. This part has a 5 V power supply, two four-bit wide ports for fast FIFO (first-in-first-out) operation, and asynchronous read and write at 33 MHz providing a bandwidth of 16MB/sec. It has cascade connection capability so two or more parts can be connected to increase storage size.

Cost is a pressing issue in consumer systems such as televisions and a smaller package with fewer leads helps reduce the cost of the part. There are no external address pins in a serial access RAM, which eliminates nine pins. The inputs and outputs are demultiplexed to improve control, which adds four pins, plus the write enable (WE\) becomes a separate Read (R) and Write (W) control. The Output Enable (OE\) is replaced by a Reset Read (RSTR) and Reset Write (RSTW) pin, resulting in the addition of one control pin, giving a total of 16 pins, so the field memory fits in a 16-pin package rather than the 20 or 20/26-pin package used by the random access 1M DRAM.

The 20/26 package is a 26-pin package with the six middle pins removed to accommodate the large early generation 1M DRAM chips. A comparison between the pinouts of the serial field memory and the standard DRAM is shown in Figure 7.1.

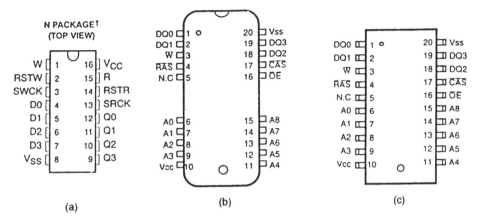

Figure 7.1 Pinout comparison of 1M serial frame memory and 1M DRAM: (a) 16-pin serial field 1M 4-bit DRAM package; (b) early 20/26-pin 1M ×4 DRAM package; (c) late generation 20-pin 1M ×4 DRAM package

Another change is that the RAS\ and CAS\ control signals are replaced with serial clocks.

The addressing is controlled by write and read address pointers which clock the data read out or written in. These must be reset to zero before a new memory access begins. The chip provides self refresh and arbitration logic to prevent conflict between memory refresh requests and data input and output cycles.

A functional block diagram is shown in Figure 7.2 which shows the DATA input and output buffers, the data cache ("A" and "B" line) buffers, the read and write counters along with the ring oscillator, the address pointers, the serial read and write timing controllers, and the read and write reset controllers.

The first 120 words of data input are stored in the "A" line cache buffer for fast access without having to access the main memory. The next data starting with word 120 goes into the 256-word write line buffer and thence into the main array. While the upper write line buffer is transferred to memory, the bottom write line buffer will be filling.

The simple frame memory storing up to 4M bits running at 33 MHz is useful for storing one frame in conventional television. High definition TV (HDTV) requires a sampling frequency of 70–80 MHz and 16Mbit of serial RAM for field memory. Two interleaved 8Mbit serial DRAMs running at 50 MHz or one 16Mbit serial DRAM running at 100 MHz can serve this application.

7.2.2 Serial DRAM for High Definition TV Frame Storage

The block diagram of an 8Mbit 50 MHz serial DRAM designed by Matsushita for HDTV applications is shown in Figure 7.3. The part has eight serial inputs and eight serial outputs. Each of the internal 128K × 8 subarrays has a serial-to-parallel and parallel-to-serial converter [10]. The bandwidth is 50MB/sec and for two of these parts interleaved is 100MB/sec.

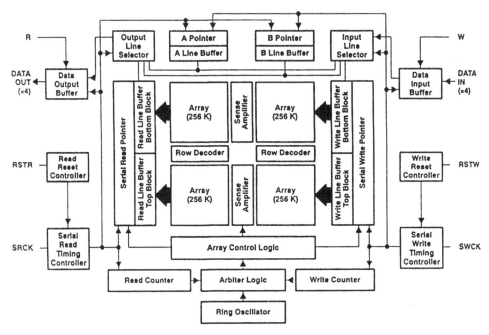

Figure 7.2 Functional block diagram of 1M field memory for TV applications (source: Texas Instruments [4])

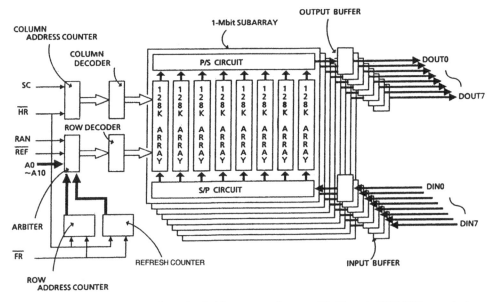

Figure 7.3 Block diagram of an 8M field memory (source: H. Kotani, 1990 [10] permission of IEEE)

A higher density DRAM can permit storage of multiple fields of video data. For example, a 256Mbit serial access DRAM, also from Matsushita, can store two seconds of NTSC data or one second of HDTV data. It has 16 serial input ports and 16 serial output ports and runs at 100 MHz providing 1.6GB/sec of bandwidth in the system.

The internal array is similar to that shown for the 8M field memory in Figure 7.3. A timing diagram of the part is shown in Figure 7.4 [11].

The serial port clock is shown running at 100 MHz. Signals are read on the rising edge of the clock. An 80 ns RAS\ latency and 40 ns CAS\ latency for read is shown.

The merger of television and computer in the new trend to multimedia systems may mean the advent of new applications-specific television memories.

7.3 Single Port Asynchronous DRAMs for Graphics Applications

Graphics subsystems in computers have moved from being only in the high end workstations and mainframes, to being almost universally present in PCs. The requirements range from fairly simple graphics in low end PC to a much higher level of complexity in mid-range to high end PCs.

7.3.1 Wide DRAM for Unified Memory in Low End PC

Low end PC systems have used wide asynchronous DRAMs to obtain the bandwidth required to implement graphics features in the system. The high bandwidth of wide

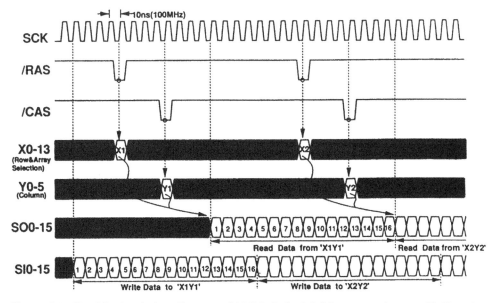

Figure 7.4 Read/write timing diagram of 256M clocked field memory (source: H. Kotani, 1994 [11] permission of IEEE)

DRAMs with fast modes such as fast page and Hyperpage (EDO) mode were described in Chapter 4. Parts such as 1M×16 DRAMs with 40 MHz Hyperpage mode cycle times are adequate for both main memory and graphics in a low end computer system.

When main memory storage and graphics memory are combined in one main memory storage location, the system is said to have a "unified memory". A unified memory using asynchronous DRAMs gives adequate bandwidth only in very low end PC systems.

7.3.2   Wide DRAM for High Speed Printer Graphics

Another system which needs high bandwidth memory for a graphics type of application is high speed printers. Printers have not required more than a few megabytes of memory storage along with 25–33 MHz speed and a wide interface. This means that the wide asynchronous DRAMs have also been used for this application.

For example, four 512K×8/9 fast page mode DRAMs running at 25 MHz on a 32-bit bus offers 100MB/sec datarate. This combination can be upgraded to a single 512K×32 25 MHz fast page mode DRAM, saving the cost of the four packages. This in turn can be upgraded in speed to a 512K×32 Hyperpage mode DRAM with a 33 MHz cycle time which gives 130MB/sec datarate [15]. New printers now in development may require considerably more memory which will still require the wide interface.

A pinout of a 28-pin 512K×8 DRAM TSOPII package is shown in Figure 7.5 along with that of a 70-pin 512K×32 DRAM TSOPII package. The savings in board space in going from four of the former to one of the latter are not as significant as in previous generations of upgrades, partially due to the number of power and ground pins which need to be added to keep the ratio of power and ground to DQs at 1:4 to control ground bounce.

7.4   Graphics Features on Asynchronous Single Port DRAMs

There has also been an attempt to add graphics features to the asynchronous single port DRAMs which are intended specifically for the graphics PC buffer applications. The write-per-bit function, also known as "mask write", is such a feature.

There is also a function called persistent write-per-bit, which allows a mask to persist for more than one cycle. These features are also implemented on the dual port graphics DRAMs to be described in a later section.

7.4.1   Write-Per-Bit

The Write-per-Bit (WPB) function [12] provides the ability to alter, or mask, some of the bits in a word while leaving other bits in the same word unaffected. If the mask is

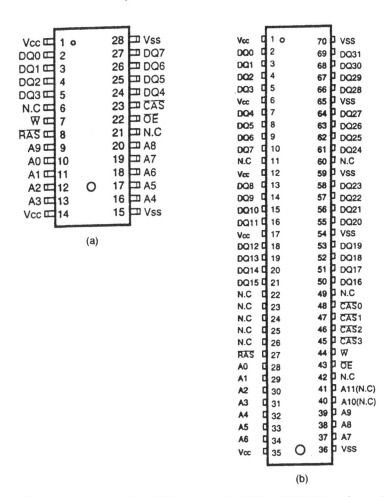

Figure 7.5   Pinout comparison of (a) 512K×8 and (b) 512K×32 DRAM packages (source: Samsung [14])

set as part of the write cycle, it can be used with no increase in cycle time over a standard read or write cycle.

Write-per-bit is implemented using a register on the data-in buffer which is latched on the falling edge of RAS\ and enabled by a low signal on the WE\ pin at a RAS high-to-low transition. A block diagram of a 256K×16 DRAM with the mask data register is shown in Figure 7.6.

This feature permits any number of bits in a word to be changed during a write cycle. In an asynchronous DRAM the mask is applied to the DQ lines and loaded into a register at the falling edge of RAS\ if the write enable (WE\) signal is low. When the DQ line is high, the corresponding bit will be written when the write cycle executes. If the DQ line is low, the bit remains unchanged.

An example of a timing diagram comparison for a simplified write-per-bit cycle and a normal write cycle is shown in Figure 7.7.

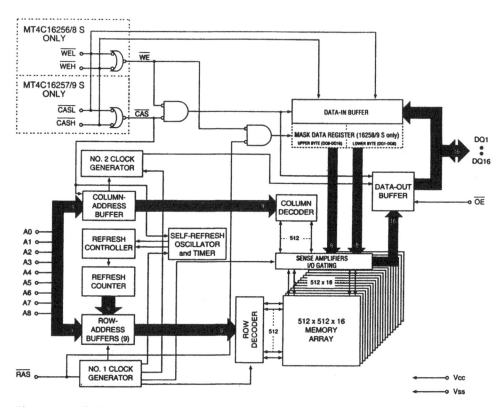

Figure 7.6   Block diagram of 256K×16 DRAM showing mask data register (source: Micron Technology [13])

Figure 7.8 illustrates the effect of the mask data on the stored data for different inputs along with the timing diagram for the masked and non-masked write for a 512K×8 DRAM [13].

During page mode operation, a mask can be loaded at the falling edge of RAS\ and will remain set and active during a write cycle as long as RAS\ remains low. A mask is effective throughout a single page mode cycle and may not be changed during this cycle as shown in Figure 7.9 [15].

7.4.2   *Persistent Write-Per-Bit*

In systems where a single mask will be used for several cycles, some chips permit a mask to be set and persist for more than one cycle. This is referred to as "persistent write-per-bit". It has not been commonly offered on the asynchronous wide DRAMs because of the additional silicon cost involved.

7.5   Synchronous Single Port DRAMs Used in Graphics Systems

The synchronous DRAMs provide additional speed for a single port DRAM up to perhaps 200 MHz. The synchronous graphics DRAMs (SGRAMs) are a functional

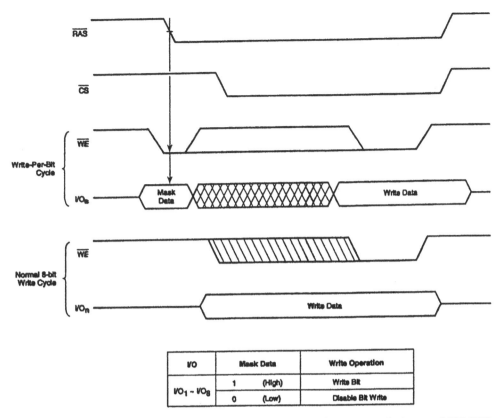

Figure 7.7   Comparison of write-per-bit cycle vs. standard 8-bit write cycle (source: NEC [12])

superset of the synchronous DRAMs. They have all of the features of the SDRAMs plus some additional features useful in graphics subsystems.

The earliest synchronous graphics DRAM was a 4Mbit part developed by a few vendors such as United Memories, Fujitsu, and Hitachi. An 8M SGRAM, which is functionally identical to the 4M, was offered by many vendors and a 16M SGRAM followed.

7.5.1   4M Synchronous Graphics DRAMs

The 4Mbit standard ×16 SDRAM is 3.3 V with an LVTTL interface and/or 5 V with a TTL interface. It is single ported and runs up to 66 MHz providing 132MB/sec of bandwidth. Graphics features supported included an eight-bit Block Write and a special function pin (DSF pin) to select between a standard SDRAM function and the SGRAM functions [17].

The block diagram of a 4M synchronous graphics RAM is shown in Figure 7.10(a) and a pinout is shown in Figure 7.10(b). The SGRAM uses two banks internally which can provide a high speed sustained burst.

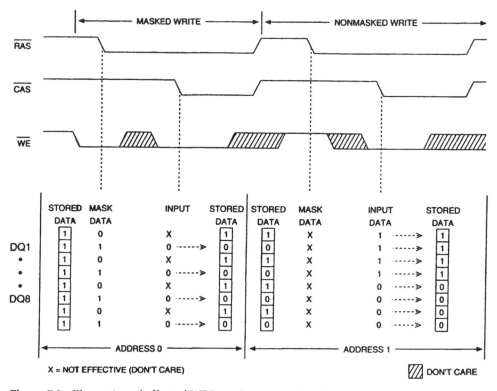

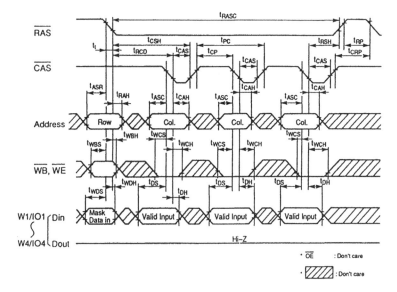

Figure 7.8 Illustration of effect of WPB mask on stored and input data (source: Micron Technology [13])

Figure 7.9 Write-per-bit during a fast page mode early write cycle (source: Hitachi [15])

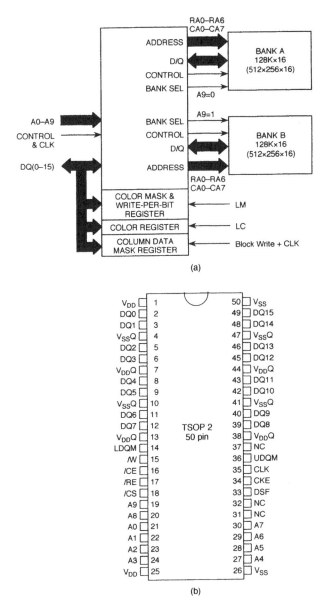

Figure 7.10 4M synchronous graphics DRAM: (a) functional block diagram; (b) pinout (source: United Memories)

The DSF pin is intended to implement the graphics features on this part. If the DSF pin is not connected, the part is intended to be a 4M SDRAM version of the standard 16M SDRAM. Graphics features include block write, and block write with auto precharge which will be described further in the next section.

7.5.2   8M SGRAMs

The 8M synchronous graphics DRAM is a standard part which is being sourced by many suppliers. It is a single port SDRAM with 3.3 V LVTTL interface. Various of the SGRAMs are also expected to offer other high speed interfaces such as HSTL or T-LVTTL. The organization is 128K (131 072) × 2 banks × 32 bits. That is, it has two banks each of which are organized 128K×32. The speed has been specified up to 100 MHz providing 400MB/sec datarate.

The standard package is the 100-pin plastic QFP with 14 × 20 mm² package dimensions and 0.65 mm pin pitch. The 8M SGRAM has a pin rotation which is similar to that of the 4M SGRAM. It is also pin compatible with the 16M SGRAM which is organized 512K×32.

The pinout and pin definitions are shown in Figure 7.11 [9].

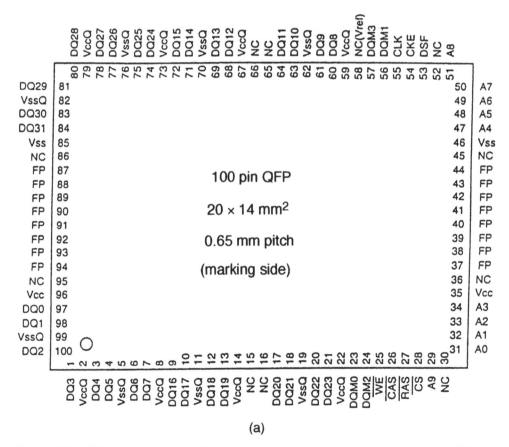

(a)

Figure 7.11   8M synchronous graphics RAM: (a) pinout; (b) pin definitions (source: NEC [9])

SYMBOL	FUNCTION
A0 – A9	ADDRESS INPUTS
A0 – A8	ROW ADDRESS INPUTS
A0 – A7	COLUMN ADDRESS INPUTS
A9	BANK SELECT
DQ0 – DQ31	DATA INPUTS / OUTPUTS
CS\	CHIP SELECT
RAS\	ROW ADDRESS STROBE
CAS\	COLUMN ADDRESS STROBE
WE\	WRITE ENABLE
DQM0 – DQM3	DQ MASK ENABLE
DSF	SPECIAL FUNCTION ENABLE
CKE	CLOCK ENABLE
CLK	SYSTEM CLOCK INPUT
VCC	SUPPLY VOLTAGE
VSS	GROUND
VCCQ	SUPPLY VOLTAGE FOR DQ
VSSQ	GROUND FOR DQ
FP	FLOATING PIN (WITH INTERNAL CONNECT TO VBB)

(b)

SOURCE: NEC [9]

Figure 7.11 *(continued)*

7.6 Special Graphics Features on SGRAMs

Special graphics features included on the 8M SGRAMs [5, 29] beyond the basic SDRAM features include masked block write, and mask write which includes the write-per-bit function. These features are standardized.

To the command functions present in the normal SDRAM mode register, the SGRAM has added Special Mode commands which control Color and Mask Registers which have also been added.

The Color Register is used in block writes and the Mask Register is used in mask writes (write-per-bit). The Mode Register with a Special Mode Register section blocked out is shown in the block diagram of an 8M SGRAM in Figure 7.12. Also shown are the Color Register and Mask Register.

The Command Truth Table of the SGRAM contains the standard command functions of the SDRAM and the Special Mode Register command functions for color and mask operation of the SGRAM as shown in Figure 7.13.

The Special Mode Register is controlled by the DSF (designated special function) control pin.

If the DSF pin is low (inactive), the 8M SGRAM operates similarly to a JEDEC Standard SDRAM. For standard SDRAM operation, the addresses A0–A8 are row addresses when the active command is given.

When CAS\ is active, addresses A0–A7 are column addresses and address A8 enables and disables the auto-precharge function. Address A9 is the bank select, BA. For BA low, Bank 0 is selected, and for BA high, Bank 1 is selected.

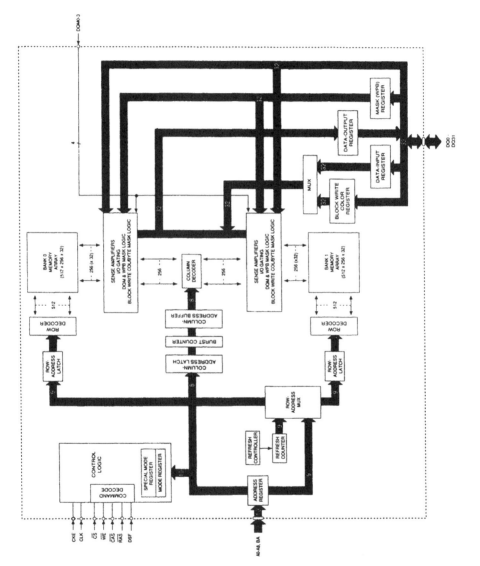

Figure 7.12   Block diagram of an 8M SGRAM (source: Micron Technology [5])

NAME (FUNCTION)	C̄S̄	R̄Ā S̄	C̄Ā S̄	W̄Ē	DSF	DQM	ADDR	DQs
COMMAND INHIBIT (NOP)	H	X	X	X	X	X	X	X
NO OPERATION (NOP)	L	H	H	H	L	X	X	X
ACTIVE (Select bank and activate row)	L	L	H	H	L	X	bank/row	X
ACTIVE with WPB (Select bank, activate row and WPB)	L	L	H	H	H	X	bank/row	X
READ (Select bank & column and start READ burst)	L	H	L	H	L	X	bank/col	X
WRITE (Select bank & column and start WRITE burst)	L	H	L	L	L	X	bank/col	VALID
BLOCK WRITE (Select bank & column and start BLOCK WRITE access)	L	H	L	L	H	X	bank/col	MASK
PRECHARGE (Deactivate row in bank or banks)	L	L	H	L	L	X	Code	X
BURST TERMINATE	L	H	H	L	L	X	X	Active
AUTO REFRESH or SELF REFRESH (enter SELF REFRESH mode)	L	L	L	H	L	X	X	X
LOAD MODE REGISTER	L	L	L	L	L	X	OpCode	X
LOAD SPECIAL MODE REGISTER	L	L	L	L	H	X	OpCode	VALID
Write enable/output enable	-	-	-	-	-	L	-	Active
Write inhibit/output High-Z	-	-	-	-	-	H	-	High-Z

Figure 7.13   Command truth table for 8M SGRAM (source: Micron Technology)

If the DSF pin is high (active), the 8M SGRAM graphics operations are active. These include the "Special Mode Register Load" cycle, and the various masked write and block write functions.

7.6.1   Load Special Mode Register Cycle

When all control pins are low and the DSF pin is high, the Special Mode Register is loaded using inputs A0–A8 and the bank select BA. The "Load Mode Register" command can be issued when both banks of the SGRAM are idle.

A block diagram of the Special Mode Register definition is shown in Figure 7.14.

7.6.2   Load Mask Register

During a "Load Special Mode Register" cycle, A5 controls the Mask Register. If A5 is "0", the Mask Register is unchanged. If A5 is "1", the Mask Register is loaded with the new data applied to the DQs. The Mask Register then acts like a per DQ bit mask during Masked Write and Masked Block Write Cycles. The mask register will retain this data until it is loaded again or until the power is turned off.

7.6.3   Load Color Register

Similarly during a "Special Mode Register Load" cycle, A6 controls the 32 bit color register. If A6 is "0" the Color Register is unchanged and if A6 is "1" and the special

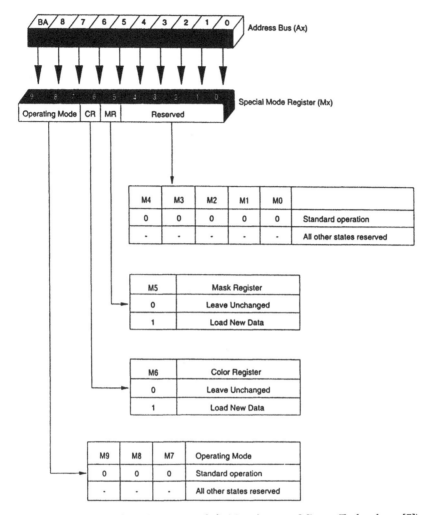

Figure 7.14 Special mode register definition (source: Micron Technology [5])

mode register load conditions are in effect, the Color Register is loaded with the data applied to the DQs. The Color Register then supplies the data during Block Write cycles. It will retain this data until reloaded or power is turned off.

The Load Special Mode Register command can be given when both banks are idle. It can also be given if either or both banks are active but no Read, Write, or Block Write access is in progress.

Other Special Mode Register inputs include A0–A4 which are all set at zero for standard operation. Other configurations of A0–A4 are reserved for future definition. A7, A8, and BA indicate the operating mode.

A truth table and timing diagram for a "Special Mode Register Load" cycle is shown in Figure 7.15.

On the rising edge of the clock, CS\, RAS\, CAS\ and WE\ are low, and DSF is high. The register data on A0–A8 and the BS is sampled and the Mask or Color

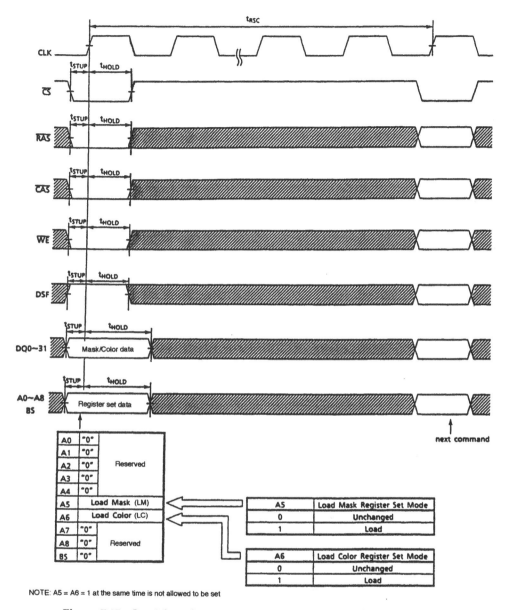

Figure 7.15 Special mode register load timing diagram (source: Toshiba)

Register data on DQ0–31 is sampled. A5 = A6 = 1 is not allowed, that is, the Color and Mask Registers can not be set on the same cycle.

7.6.4 *Active Graphics Commands*

Commands with DSF high are also shown in the Command Truth Table for activate Masked Write (Write-per-Bit) and activate Block Write. Otherwise the DSF pin is a "don't care", although some manufacturers recommend that it be held low for compatibility with future, as yet undefined, special modes.

7.6.5 Masked Write-Per-Bit

The "Active with Mask Write (Write-per-Bit)" command is similar to an active command during which the write-per-bit is activated. Any subsequent write or block write cycles to the selected bank or row will be masked in accordance with the contents of the 32-bit Mask Register, the DQM signals, and, in the case of a Block Write command, the column/byte mask information from the color register [5].

The write-per-bit data acts as an I/O (DQ) mask. It uses the bits in the Mask Register to mask various data inputs applied to the DQ pins during the write cycle as illustrated in Figure 7.16. This figure uses for illustration only the first I/O (DQ) byte (first eight bits of the 32-bit I/O).

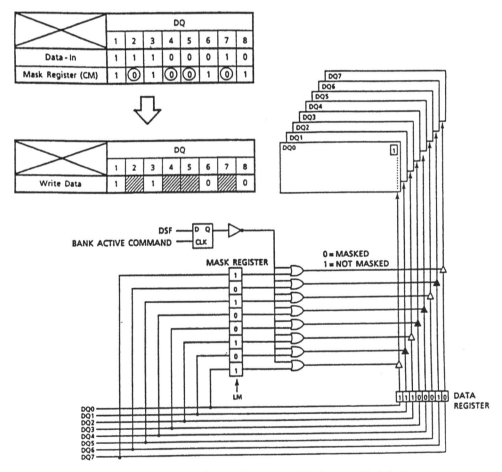

Figure 7.16 Masked write (write-per-bit) (source: Toshiba)

Write-per-bit can be burst. If DSF is low when the Bank Active command is executed, then the write-per-bit mask will be disabled for that cycle. If DSF is high at the Bank Active command the mask will be enabled and the DQ inputs will be masked.

7.6.6 Block Write

The "Block write" command is used to write a single 32-bit word into a block of eight consecutive column locations in one row. These column locations are designated by a starting column addresses (A3C–A7C) and the bank select. The single 32-bit data value comes from the color register which must have been previously loaded. This is illustrated in Figure 7.17.

In this figure we see that a "1" from the least significant bit of the color register (LC) is stored in each of the eight column locations in the DQ0 array. Similarly a "0" from the next bit of the color register is stored in the DQ1 array, etc.

The data from the color register is masked by the data from the mask register where "1" in the mask register enables the data from the color register and "0" in the mask register disables it. For example the "0" in the LM+2 location of the mask register disables the "1" applied from the LC+2 location of the color register so that the column-byte in DQ2 has a "0" written as shown in Figure 7.16.

The data on the DQs when the Block Write command occurs can be used to mask specific column-byte combinations within the block. DQM signals are applied to all eight columns for each byte DQM (0–3). This is sometimes referred to as a "column data mask". For example, in Figure 7.16, the CM+1 bit of the column of data stored in DQ0 is masked by the column data mask so the "1" written here from the color register does not appear.

Figure 7.18 illustrates the effect of the Mask Register, Color Register and Column Data Mask for columns 0–7, the lower byte.

A timing diagram for a page mode block write for CAS Latency 3 is shown in Figure 7.19. In cycle 1, Bank 1 in activated. In cycle 3, a Block Write is activated for Bank 1. The Bank Select pin (BS) is high indicating bank selected and A8 is low indicating Bank 1. DSF is high activating the color and mask registers for the I/O (DQ). Column addresses A3–A7 select the first block of eight columns to be written out of 32 blocks in the row. The DQM provide the mask for the column byte input on the DQ.

7.6.7 Clock Enable on SGRAM

The Clock Enable (CKE) truth table is unchanged from the SDRAM. It indicates the logic state of CKE at clock edge n and the state at the immediately prior state $n-1$. This truth table was shown previously in Figure 6.24.

7.6.8 Current State Truth Table on SGRAM

While the Current State Truth Table was included in the Command Truth Table, it is broken out in Figure 7.20 for closer examination. The current state is the state of the SGRAM immediately prior to clock edge n.

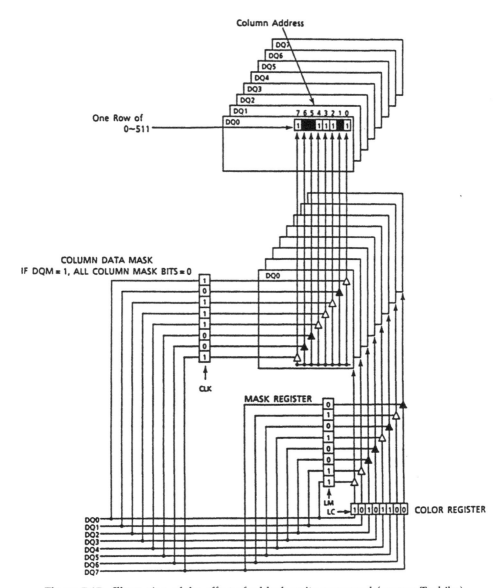

Figure 7.17 Illustration of the effect of a block write command (source: Toshiba)

The current state describes the commands issued to a specific bank. An idle bank is one that has been precharged with t_{RP} met. "Row Active" means the bank has been activated and t_{RCD} met, but no accesses are in progress. "Read" or "Write" means a read or write burst has been initiated but has not yet terminated.

The time for the underlying DRAM to perform the action initiated must be met in all cases. For example, "Precharge" starts with the clocking in of the precharge command and ends after time t_{RP}. So when activating a row, t_{RCD} must be met, and when refreshing t_{RP} must be met. The "Burst Terminate" command is not bank specific but affects any Read or Write burst in progress in either bank.

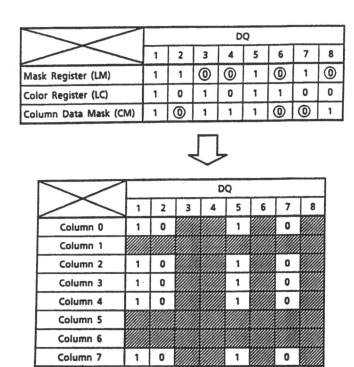

		DQ							
		1	2	3	4	5	6	7	8
Mask Register (LM)		1	1	⓪	⓪	1	⓪	1	⓪
Color Register (LC)		1	0	1	0	1	1	0	0
Column Data Mask (CM)		1	⓪	1	1	1	⓪	⓪	1

	DQ							
	1	2	3	4	5	6	7	8
Column 0	1	0			1		0	
Column 1								
Column 2	1	0			1		0	
Column 3	1·	0			1		0	
Column 4	1	0			1		0	
Column 5								
Column 6								
Column 7	1	0			1		0	

Figure 7.18   Lower byte mask/color registers and column mask (source: Toshiba)

7.7   16M SGRAMs

Currently 16M SGRAMs are available. Graphics subsystems in PCs have increased in density requirements to 2–4 MB for a notebook computer and 4 MB and higher for a desktop PC. The 16M SGRAMs have retained the 32-bit interface and are available with the LVTTL interface and also with faster interfaces such as the SSTL3 and SSTL2. The 16M SGRAM continues to be offered in the 100-pin QFP package with a tighter pin pitch.

A graphics system requiring 2MB of graphics memory can use one 16M SGRAM on a 32-bit bus, while a system requiring 4MB of graphics memory can use two 16M SGRAMs on a 64-bit bus.

The basic features of the SGRAM did not change between the 8Mb and the 16Mb devices. Many of the SGRAMs have been offered without the additional graphics features such as masked Write and Block Write since 32-bit wide synchronous DRAMs were also used for other applications requiring high bandwidth and low granularity in areas such as computer peripherals and networking. It is also possible to use a narrower bus SDRAM in a graphics application by implementing the graphics features in the controller rather than in the DRAM.

The higher speed SGRAMs are offered with the SSTL interface. An example is a 512K×32 SGRAM offered by MoSys Inc. This part uses a higher speed DRAM core

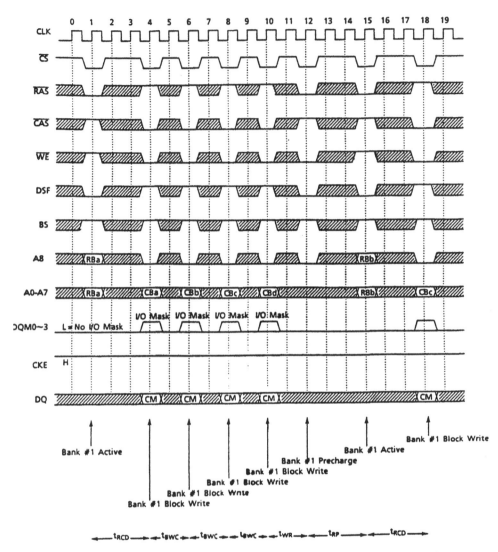

Figure 7.19   Timing diagram for a page mode block write (source: Toshiba)

which is described in a later section of this chapter. Operation up to 200 MHz is possible. Two or four independent banks are available.

7.8   Other Single Port Graphics DRAMs

Various other single port DRAMs which are primarily targeted at graphics subsystems have been announced. They include parts from Mosys, Rambus, Hitachi, Mitsubishi, and Neomagic among others. Brief descriptions of these alternative single port DRAMs follow.

CURRENT STATE	CS	RAS	CAS	WE	DSF	COMMAND/ACTION
Any	H	X	X	X	X	COMMAND INHIBIT (NOP/ continue previous operation)
	L	H	H	H	L	NO OPERATION (NOP/ continue previous operation)
	L	L	H	H	L	ACTIVE (Select bank and activate row)
	L	L	H	H	H	ACTIVE w/WPB (Select bank, activate row and WPB)
Idle	L	L	L	H	L	AUTO REFRESH
	L	L	L	L	L	LOAD MODE REGISTER
	L	L	L	L	H	LOAD SPECIAL MODE REGISTER
	L	H	L	H	L	READ (Select bank and column and start READ burst)
	L	H	L	L	L	WRITE (Select bank and column and start WRITE burst)
Row Active	L	H	L	L	H	BLOCK WRITE (Select bank & column and start BLOCK WRITE access)
	L	L	H	L	L	PRECHARGE (Deactivate row in bank or banks)
	L	L	L	L	H	LOAD SPECIAL MODE REGISTER
	L	H	L	H	L	READ (Select bank and column and start new READ burst)
READ	L	H	L	L	L	WRITE (Select bank and column and start WRITE burst)
(AUTO- PRECHARGE	L	H	L	L	H	BLOCK WRITE (Select bank & column and start BLOCK WRITE access)
DISABLED)	L	L	H	L	L	PRECHARGE (Truncate READ burst, start precharge)
	L	H	H	L	L	BURST TERMINATE
	L	H	L	H	L	READ (Select bank and column and start READ burst)
WRITE	L	H	L	L	L	WRITE (Select bank and column and start new WRITE burst)
(AUTO- PRECHARGE	L	H	L	L	H	BLOCK WRITE (Select bank & column and start BLOCK WRITE access)
DISABLED)	L	L	H	L	L	PRECHARGE (Truncate WRITE burst, start precharge)
	L	H	H	L	L	BURST TERMINATE

Figure 7.20   Current state truth table (source: Micron Technology)

7.8.1   *Synchronous Protocol DRAM from Rambus, Inc.*

A synchronous DRAM type developed by Rambus, Inc. has been made in various densities from 4Mb to 18Mb.

This DRAM has a proprietary interface which permits data transfer on a ×8 or ×9 wire bus at a peak transfer rate of 500MB/sec per DRAM [7]. Multiple DRAMs can be on a single bus [19]. Each acts as a slave in responding to bus transactions initiated by a proprietary master device which is a specialized ASIC. The parts are in vertical single-in-line or in surface mount packages which can be lined up on a bus, or "channel". The channel consists of 29 wires, as shown in Figure 7.21 including 8 or 9 data, 8 ground, 5 power, 2 clock, a VREF and various control pins [28].

The "master" controller chip sends out request packets to all the "slave" DRAMs on a single bus channel. All the DRAMs on a channel receive the request packet which contains the address and control information specifying the requested operation. Each DRAM examines the packet. The one which is addressed sends back either an "acknowledge" or a "busy" signal, called a NACK, to the controller.

If the DRAM which is addressed is busy, then it sends back a "NACK" signal to the controller and the controller tries the access again later. If the addressed DRAM is not busy, it returns an acknowledge signal then, the read or write transaction that was requested proceeds.

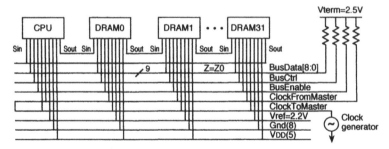

Figure 7.21   Illustration of the configuration of a 4.5Mb 500Mb/sec protocol DRAM (source: N. Kushlyama (1993) [28] with permission of IEEE)

If the DRAM is available and a write was requested, the master follows the request packet with a data packet which is written into the sense amplifiers if the row corresponding to that data is currently on the sense amplifiers. This is called a write hit and is shown in the timing diagram in Figure 7.22(c) [28].

If the DRAM is available, a read is requested, and the data stored at the address requested is present on the sense amplifiers; this is called a "read hit" and is shown in the timing diagram in Figure 7.22(a). The DRAM sends the requested data on to the master over the special channel. The master then converts the high speed data on the channel back to the standard interface used in the system. In cases where the DRAM is available and there is a "hit" the datarate can be as high as 500MB/sec for the burst of data accessed. This figure is for the burst and does not allow for bus protocol overhead and initial latency.

If the DRAM is available when a read or write is requested but the correct row address is not currently on the sense amplifiers then there is a "miss". The DRAM proceeds to access the addressed row and when the data is on the sense amplifier either sends it to the master (in the case of a read), or writes the data packet to the sense amplifier (in the case of a write). Figure 7.22(b) shows the case of a DRAM available with a read miss and access, and Figure 7.22(d) shows the case of a DRAM available with a write miss and access. The penalty for a miss is the need to incur the RAS\ access time of the DRAM to access the requested row. For misses, initial access was 152 ns for a read or 120 ns for a write for the early 18Mb density part [19].

In the case where the DRAM is not available when first requested and then a miss occurs when it is available, the resulting access time can be significantly delayed.

Since standard DRAM technology underlies the part, there is an initial latency before the peak transfer rate is achieved during which the various DRAMs are accessed and the sense amplifiers filled.

The 500MB/sec peak transfer rate during a burst is equivalent to that which could be achieved with four 60 MHz 1M×16 SDRAMs on a 64-bit bus, or two 120 MHz 1M×16 SDRAMs on a 3-bit bus. For the same datarate, smaller granularity is therefore possible with this DRAM.

This DRAM has a proprietary synchronous protocol for fast block-oriented transfers and a proprietary low signal swing interface to the proprietary bus. There are on-chip registers which permit flexible addressing control. There are two cache lines per

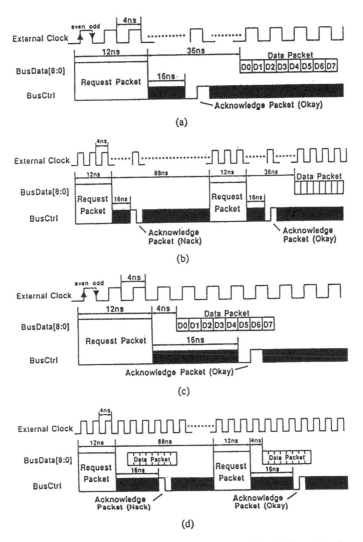

Figure 7.22 Timing diagrams for the 4.5Mb synchronous DRAM from Rambus, Inc.: (a) read hit; (b) read miss; (c) write hit; (d) write miss (source: N. Kushlyama (1993) [28] with permission of IEEE)

DRAM with each cache line being 1KB each. The 18M DRAM has on-chip RAS\, CAS\ and refresh logic. It is packaged in a proprietary vertically oriented 32-pin single-in-line surface mount plastic package [7].

New pins on the chip include the bus data pins for request, write and read protocols. The data lines, which are active low, carry the request packet with the address, operation codes and the count of the bytes to be transferred. The receive clock (RxClk) is aligned with incoming request and write data packets. The transmit clock (TxClk) is aligned with the data being sent out on reads and in acknowledge packets. The reference voltage (V_{Ref}) is the logic threshold voltage for the low swing signals. The bus control is a control signal to frame packets, transmit opcodes and to acknowledge requests. The bus enable is a control signal which is pulsed to power-up the bus. The daisy chain input (S_{In}) and output (S_{Out}) pins are used to reset the daisy chain input and output. They are active high [7].

A graphics controller for this DRAM has been designed by Cirrus Logic and may result in this part being used in some 1MB frame buffer graphics applications in PCs [20].

7.8.2 Multiple Bank DRAM from Mosys

On all of the DRAMs considered to this point, latency for the first access has been a problem. Column access time for data already on the sense amplifiers of the single port DRAMs has been getting faster through the use of wider, faster interfaces and faster access modes.

The SDRAM and the Rambus, Inc. DRAM introduced the concept of reducing the latency for the first access by interleaving two banks. If both banks already have a row of data active, and the burst that the processor wants next is on one of these rows, then there is no delay. If the requested row is not active, then the latency for the first access may be only partially hidden.

The addition of graphics features integrated into the DRAM has provided some help with the graphics operations requiring random bit accessability.

A DRAM from Mosys, Inc. has attempted to solve these problems by integrating many small DRAM banks onto a single chip all connected to a fast common bus internal to the chip and controlled on chip for clock skew.

The DRAM RAS\ latency problem is reduced by the small size of the DRAM banks and the column latency is reduced by the very short CAS lines to the bus. A schematic block diagram is shown in Figure 7.23.

The interface is synchronous. All signals enter and exit the chip at the control end.

Speed is gained because a small DRAM is inherently faster than a larger DRAM in the same technology due to reduced wordline capacitance and shorter internal wiring. For example, the organization of the banks of the ×32 Mosys DRAM is 256 × 32 × 32 compared with the organization of a 16M ×16 DRAM which is 1024 × 1024 × 16. The 256 × 32 banks used in the Mosys DRAM in a 16M technology are individually faster than 2048 x 1024 bank of the 16M DRAM.

The RAS access time of this multi-bank DRAM in a 0.5 micron technology is 36 ns, CAS access time is 12 ns and there is a 6 ns burst cycle time (166 MHz). The external bus is 16 bits wide. Both clock edges are used for synchronous transfer of 32 bits per clock cycle, so peak bandwidth for burst accesses is 668MB/sec.

The speed is maintained by using a fast metal bus across the chip. Clock skew is avoided on the chip by having the internal bus terminated on the I/O side of the chip. Byte level write mask capability is provided.

Not only are the individual banks faster, but the existence of more than the two banks present in the 16M SGRAM and the Rambus, Inc. DRAM increases the probability of a hit on an accessed row, thereby reducing the frequency of occurrence of time-consuming row accesses. This increases the average bandwidth of the chip.

ADDRESS, I/O, CLOCK, CONTROL

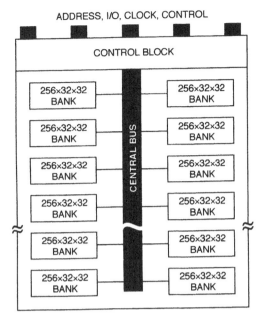

Figure 7.23 Block diagram of Mosys DRAM chip concept (source: Mosys)

Improvements in average bandwidth of the multibank DRAM over the two-bank SDRAM and the one-bank EDO DRAM are shown in Figure 7.24.

The Mosys DRAM up to 576K×32 is offered in a 128-pin plastic QFP or a 50-pin PLCC. A 200 MHz SGRAM has been introduced using this technology.

Mosys attempted to minimize the problem that all applications-specific DRAMs have – increased silicon area and hence cost.

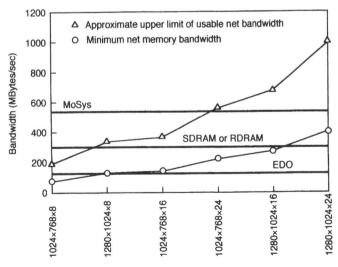

Figure 7.24 System bandwidth requirement and minimum net bandwidth by DRAM type (source: Mosys)

First, they noted that the various frame buffer applications always waste memory since frame resolutions and DRAM densities never quite match. For example a 1024×768×16 screen resolution requires 12.39Mbits of memory storage. If a 16M DRAM is used then 3.71Mbits of memory are wasted, that is, the 16M is 25 percent too large for this application.

If a Mosys DRAM with 50 256K banks is used, the amount of data storage is 12.8M. This is enough to service the basic application and still permit the chip to remain within the chip size of the 16M DRAM. One drawback is that graphics subsystems designers frequently find a use for the additional memory.

Secondly, Mosys noted that if a few extra banks are included, then yield can be increased by substituting good redundant banks for defective banks, thereby increasing the overall chip yield. A drawback is the chip size must be larger by the size of the redundant banks. This can be a compelling argument in the early years of a high density DRAM when the production yields are low. It may be less so when the technology is mature and the basic yields are high.

The main benefits of this part are expected to be low pin count, minimum memory configuration, and the low latency.

Notice in Figure 7.24 that even the bandwidth of the Mosys DRAM is not claimed to service frame buffers with resolutions of 1280 × 1024 × 16 and higher. These higher end systems fall into the workstation category and have historically been serviced by the multiport video DRAM (VRAM) or by large banks of interleaved DRAMs.

The requirement for graphics buffers with bandwidth beyond that possible with single port DRAMs still exists. It appears to be moving in two directions, those systems continuing to use multiport DRAMs, and those considering applications-specific memories.

In the next section multiport DRAMs are discussed, and in the following section some of the new applications-specific DRAMs now appearing are discussed.

7.9   Overview of Multi-Port Graphics DRAMs (VRAMS)

VRAMs attempt to compensate for the slow speed of the DRAM in video display applications by having both a random port and a serial port. The random port operates like a conventional DRAM. The serial port provides continuous refresh to a raster display screen. High bandwidth transport of information occurs between the two memories. A part with dissimilar ports is called multi-ported.

There are three basic VRAM operations. These are asynchronous random read and write access of the parallel RAM port, high speed clocked access of the serial SAM port, and transfer of data between any row in the RAM and the SAM (Serial Access Memory).

The RAM and SAM ports can be independently accessed at any time except during an internal transfer between the two memories. Some VRAMs provide bidirectional transfer of data between the RAM and the SAM, while others provide only for transfer from the RAM to the SAM. All VRAMs have high speed serial read capability and some have serial write capability.

Since the high end graphics market is small, the number of DRAM suppliers who make VRAMs is small. At the 4M density there are at least three suppliers of VRAMs including TI, IBM, and Toshiba.

7.10   An Introduction to VRAMs, the 4M VRAMs

A block diagram of a typical 4M VDRAM [21] which is organized 256K × 16 is shown in Figure 7.25.

Shown are the 256K × 16 RAM and upper and lower transfer gates of the split 256 × 16 Serial Data Register (SAM). The RAM array is organized 512 × 512 × 16. Extended page mode, which is similar to Hyperpage (EDO) mode, is used for RAM accesses. The SAM has serial read and write capability and data can be transferred between any row in the RAM and the SAM. Each half of the split SAM register is 128 × 16 bits. Mask operations are supported by a write-per-bit function, a color register, block write control logic, and flash write control logic.

A pinout of a 4M VDRAM in a TSOPII and SOG (SSOP) package is shown in Figure 7.26.

The DQ pins are the RAM I/O's, the SQ pins are the SAM outputs, the SC pin is the serial clock and the SE\ is the serial enable.

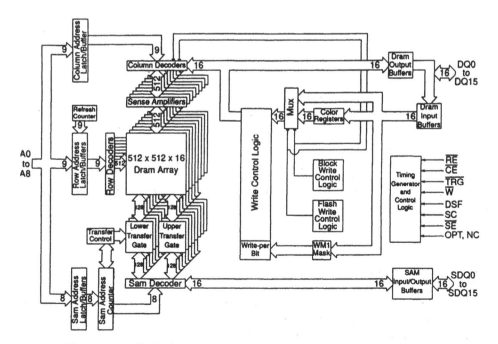

Figure 7.25   Block diagram of a typical 4M VDRAM (source: IBM [21])

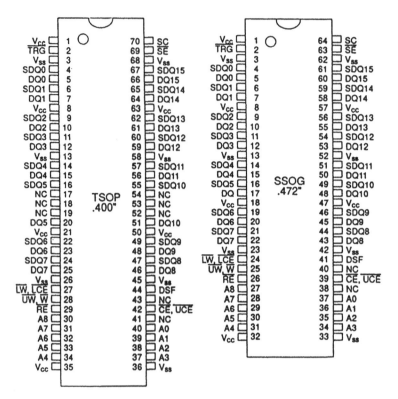

Figure 7.26   Pinout of a typical 4M VDRAM (source: IBM [21])

A TRG\ pin selects either the data transfer operation or the DRAM operation. During DRAM operation, the TRG\ is held high as RAS\ falls and acts as an output enable for the RAM DQ pins. For transfer the TRG\ is held low.

The WE\ pin enables the "write-per-bit" or mask function when it is held low as RAS\ falls, and acts as the Write Enable if held high as RAS\ falls.

The DSF pin is the special function control for the CBR refresh, the split register transfer, and the various mask functions. Mask functions include block write, non-persistent write-per-bit, persistent write-per-bit, and load color register.

The DSF pin is also a special flag output pin that indicates which half of the SAM is being accessed. When DSF is low the serial address pointer is accessing the least significant 128 bits of the SAM and when DSF is high it is accessing the most significant 128 bits.

7.11   RAM Operations

Asynchronous RAM port operations include normal random access read, early and late write, extended page mode read and write, read–modify–write, load mask register, and load color register and flash write.

Figure 7.27 shows a functional table for random access operation.

Menu Code	$\overline{RE}$				$\overline{CE}$	Address		DQ$_i$			FUNCTION
	$\overline{CE}$	TRG	$\overline{W}$	DSF	DSF	$\overline{RE}$	$\overline{CE}$	$\overline{RE}$	$\overline{CE}$	$\overline{CE}$, $\overline{W}$	
CBR	0(5)	X	1(4)	0	-	X	-	X	-	-	$\overline{CE}$ before $\overline{RE}$ Refresh
CBRS	0(5)	X	0(3)	1	-	Stop	-	X	-	-	$\overline{CE}$ Before $\overline{RE}$ Refresh and mode set (2)
CBRN	0(5)	X	1(4)	1	-	X	-	X	-	-	$\overline{CE}$ Before $\overline{RE}$ Refresh without mode reset
ROR	1	1	1	X	-	Row(1)	-	X	-	-	$\overline{RE}$ Only Refresh
LCR	1	1	1(4)	1	1	Row(1)	X	X	X	Color	Load Color Register
LMR	1	1	1(4)	1	0	Row(1)	X	X	X	Mask	Load Mask Register
RT	1	0	1(4)	0	X	Row	TAP	X	X	X	Read Transfer
MWT	1	0	0(3)	0	X	Row	TAP	WPBM	X	0	Write Transfer (Masked)
SRT	1	0	1(4)	1	X	Row	TAP	X	X	X	Split Read Transfer
MSWT	1	0	0(3)	1	X	Row	TAP	WPBM	X	X	Split Write Transfer (Masked)
RW	1	1	1(4)	0	0	Row	COL	X	X·	Data Input	Read Write Cycle (No Mask)
RWM	1	1	0(3)	0	0	Row	COL	WPBM	X	Data Input	Read Write Cycle (Masked)
BW	1	1	1(4)	0	1	Row	COL A3-A8	X	-	ADDR mask	Block Write Cycle (No Mask)
BWM	1	1	0(3)	0	1	Row	COL A3-A8	WPBM	X	ADDR mask	Block Write Cycle (Masked)
FWM	1	1	0(3)	1	X	Row	X	WPBM	X	X	Flash Write Cycle (Masked)

Notes:

1. Row address needed only for refresh operation to the selected row. Otherwise this is a don't care.
2. This cycle is used to put the chip into special modes. The A$_i$ at $\overline{RE}$ fall select the desired mode.
3. Either $\overline{W}$ is 0.
4. Both $\overline{W}$ are 1.
5. Either $\overline{CE}$ is 0 on Dual CE parts.

Figure 7.27   4M VRAM function table for RAM operation (source: IBM [21])

7.11.1   Extended Read and Write Mode

The extended read and write mode functions are similar to the Hyperpage mode on the single port RAM in that the rising edge of CAS\ no longer controls the outputs during a burst. This is illustrated in the timing diagram for fast page mode read cycle shown in Figure 7.28.

4M VRAMs have extended read and write mode functions on the RAM port that run as fast as 25 ns cycle time [21].

7.11.2   Random Port Mask Functions

The mask functions are similar to those discussed for the wide asynchronous DRAMs with graphics features. The difference between non-persistent write-per-bit and persistent write-per-bit follows.

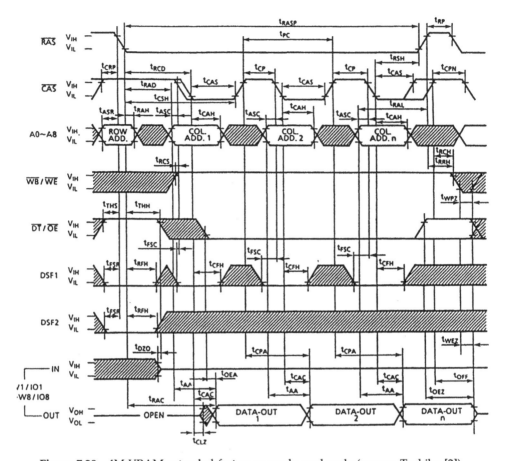

Figure 7.28   4M VRAM extended fast page mode read cycle (source: Toshiba [2])

Non-persistent write-per-bit (also called "new mask mode") permits a mask to be loaded via the 16 DQs on the falling edge of RAS\ on any write cycle. After the mask is latched, valid input on the DQ pins will be masked for that cycle only.

Persistent write-per-bit (also called "old mask mode") requires a "load mask register" cycle to load the mask into the internal mask register. Valid input data is then masked from the internal register and any inputs on the DQ pins on the falling edge of RAS\ are ignored. A CBR\ refresh cycle is required to reset this cycle.

The block write is also similar to that discussed for the single port DRAMs. Data from the color register is used to write up to 64 bits of data simultaneously to one block of the memory array. The block is implemented as 4 columns × 4 DQs for some 4M VRAMs and as 8 columns × 8 DQs or 8 × 16 for others.

Either persistent or non-persistent write-per-bit can be applied to the block write cycle.

Figure 7.29 shows a timing diagram for a load-color-register cycle, followed by a block-write cycle with no write mask, and a block-write cycle where the mask is loaded and used [22].

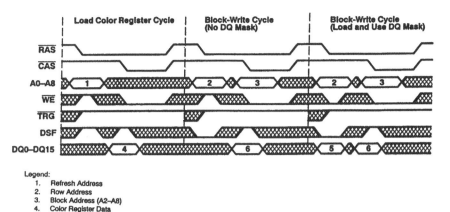

Legend:
1. Refresh Address
2. Row Address
3. Block Address (A2–A8)
4. Color Register Data
5. DQ Mask Data: DQ0–DQ15 are latched on $\overline{RAS}$ falling edge.
6. Column Mask Data: DQ_i–DQ_{i+3} (i=0,4,8,12) are latched on either the first $\overline{WE}$ falling edge or the falling edge of $\overline{CAS}$, whichever occurs later.

▧▧▧ = Don't Care

Figure 7.29   4M VRAM timing diagram showing a masked block-write cycle (source: Texas Instruments [22])

7.11.3 Flash Write

Flash Write is available on some of the 4M VRAMs. It is a special RAM cycle which lets the data in the color register be written into all the memory locations of a selected row. Each bit of the color register corresponds to one of the RAM I/O blocks, so a Flash Write cycle writes to a plane of the RAM array corresponding all the I/O locations in a single row. It can be masked by a write-per-bit function on an I/O basis. Figure 7.30(a) illustrates schematically an array plane written by a single masked Flash Write cycle (lower byte only) and Figure 7.30(b) shows a simplified Flash Write timing diagram.

Flash Write is used for fast plane clear operations. If a different row address is specified for each Flash Write cycle, the entire array can be cleared in 512 Flash Write cycles. Assuming a cycle time of 130 ns, a plane clear operation can be completed in less than 66.6 microseconds.

7.12   Transfer Operations between the RAM and SAM

For transfer operations between the RAM and the SAM, there are two cases, one for the 256 × 16 SAM, and one for the 512 × 16 SAM.

7.12.1   256 × 16 SAM

The SAM illustrated in the block diagram in Figure 7.25 is 256 × 16 bits and one row of the RAM array is 512 × 16 bits. During a transfer operation from the RAM to the SAM, the 256 × 16 bits from half of a 512 × 16 bit row in the RAM are transferred to the SAM.

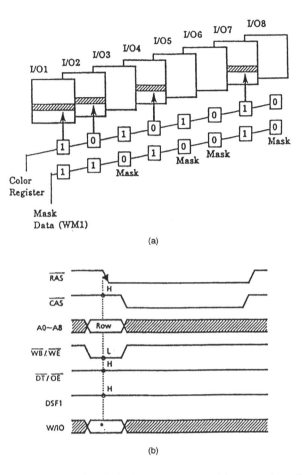

(a)

(b)

Figure 7.30   Illustration of a single Flash Write operation: (a) array plane for single masked Flash Write cycle; (b) simplified Flash Write timing diagram (source: Toshiba [2])

The transfer can be done either as a full register operation or as a split register operation.

For a full register transfer read operation, nine row addresses (A0–A8) are latched at the falling edge of RAS to select one of the 512 rows for transfer. Address A8 selects which half of the row is to be transferred. A0–A7 select one of the SAM's 256 available tap points from which the serial data is read out. A full register transfer read is illustrated in Figure 7.31(a).

In a split register transfer read operation, the serial data register is split in half as illustrated in Figure 7.32(b). The lower half contains bits 0 to 127 and the upper half contains bits 128 to 255. The advantage of the split register operation is that while one half of the SAM data is being read out of the SAM port, the other half can be loaded from the memory array. Split register transfer provides the means for continuous loading of the SAM without interrupting the continuous flow of

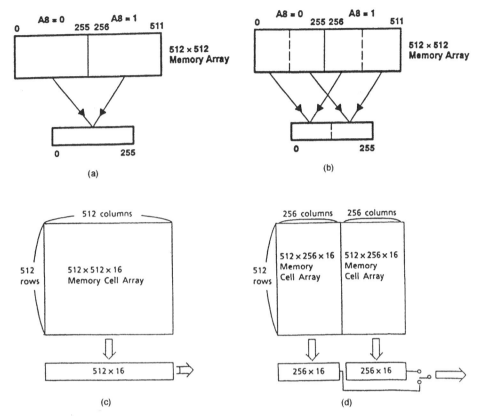

Figure 7.31   Register transfer read (RAM to SAM): (a) 256-bit SAM – full register transfer (b) 256-bit SAM – split register transfer (source: TI [18]; (c) 512-bit SAM – full register transfer; (d) 512-bit SAM – split register transfer (source: Toshiba [2])

data from the SAM port by alternating the halves of the SAM that are being loaded and read.

7.12.2   512 × 16 SAM

The 4M VRAMs having a 512 × 16 SAM load an entire 512 × 16 row of the RAM into the SAM during a transfer read operation as illustrated in Figure 7.32(c). A split register transfer read in this case is from one half of the RAM row to one half of the SAM register as ilustrated in Figure 7.32(d).

Figure 7.32 shows the timing diagram for split register transfer read timing for a 256 × 16-bit SAM with Tap Point "M" indicated by addresses A0–A6, A7 is "don't care", and A8 identifies the specified half of the row.

Pointer control permits definition of the starting locations of the serial port in a multiport DRAM and simplifies the control logic required for scrolling and hardware windows. The tap pointer is a counter that defines the starting point in the serial data

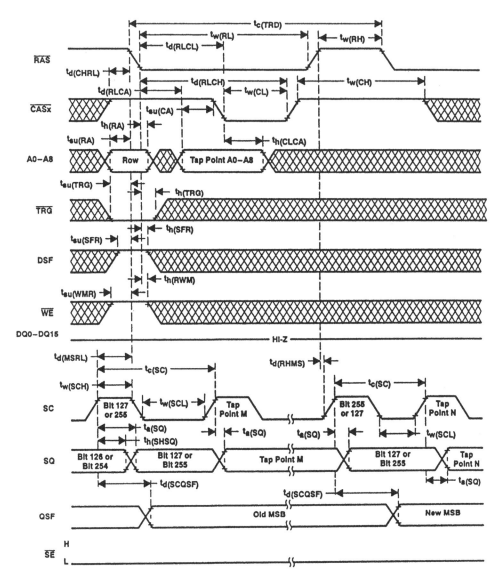

Figure 7.32   Split register transfer read timing (source: Texas Instruments [18])

register into which data is entered. The data is entered serially and wraps around when the end of the register is reached.

7.13   SAM Operation

Data is shifted out of the SAM registers at the rising edge of the serial clock (SC) with the Serial Enable (SE\) held low.

The timing diagram for a simple serial read cycle is shown in Figure 7.33.

The TRG\ is "don't care" except when RAS\ falls when it must be high to avoid initiation of a register-data transfer operation.

The SAM of a 4M VRAM can run as fast as 66 MHz [21].

7.14   Video DRAM Standards and Market

The JEDEC JC42.3 Standards Committee has specified mandatory and optional features for the 2M and 4M VRAM.

Features which are required on all 2M and 4M VDRAMs which claim to be JEDEC standard include read transfer cycle, split read transfer cycle, read write cycle (non-mask), read write cycle with new mask data, and CAS\ before RAS\ refresh.

Mandatory optional features of which, if any of the features are used, all must be used, include CBR refresh/stop, mask write transfer, block write with no mask, block write cycle, flash write with mask, and load color register.

Beyond the mandatory optional features, there are optional features which are at the discretion of the manufacturer. These include split write transfer, load mask register, pipelined fast page mode and extended out fast page mode, and a logic function set cycle.

The major drawback of the multiport RAMs has been the additional silicon, and hence cost, they incur over the silicon of the standard DRAM. The VRAM size has ranged up to 40 percent over that of a comparable DRAM. The additional features were a major part of this size increase.

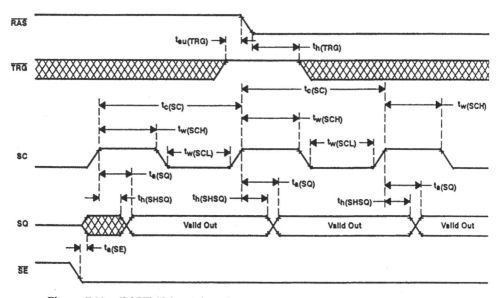

Figure 7.33   4M VRAM serial read operation (source: Texas Instruments [18])

In response to the competition from high bandwidth single port DRAMs, VRAMs have been simplified so that the silicon adder is now normally between 7 and 15 percent. An example of such simplification is going from a 512-bit to a 256-bit SAM. At a lower cost the VRAMs still provide higher bandwidth in a high end frame buffer environment than a single port DRAM.

7.15   8M Video DRAM

7.15.1   Samsung "Window RAM"

An example of a simplifed VRAM is the 8M VRAM from Samsung [23] which they call the "Window RAM". This has been the only 8M density VRAM to appear to date. A block diagram is shown in Figure 7.34.

The 256K×32 RAM consists of 32 I/O blocks of 512 × 512 arrays. The split SAM is two small 64 × 16 SAMs in parallel with 2:1 multiplexed 16-bit serial output for speed.

A minimum number of features are included. Mask function is handled with an eight column block write which allows a 32-byte block of data to be transferred at one time. There is a two-bit color register and bit and byte mask capability. Scroll and "aligned block move" operations, also known as bit block transfer (BitBLT), are done with four 256-bit latches which are used as temporary storage for internal DRAM read cycles. Serial read with split register read transfer is provided. A drawback of this part is that the display refresh overhead is higher than other VRAMs due to the narrower 256-bit load path to SAM.

The part is packaged in a 120-pin PQFP. The pinout is shown in Figure 7.35.

The four Byte Enable (BE\) pins select I/O bytes with BE0 corresponding to DQ0–DQ7, BE1 to DQ8–DQ15, etc. BE\ high disables the I/O byte, and BE\ low disables it. Color register selection is made by $D(i)$ with $D(i) = 1$ selecting color register 0 and $D(i) = 2$ selecting color register 2. Simplified RAS\ and CAS\ control truth tables are shown in Figure 7.36.

An output enable (OE\) control permits control of the data out during extended data out read and write cycles since the rising edge of CAS\ no longer controls the I/O. During a new row access the state of OE\ when RAS\ falls controls mask register update: if OE\ $= 1$ the mask register content is updated, if OE\$=0$ the cycle uses previously loaded mask data. A read–write cycle timing diagram is shown in Figure 7.37.

The color and mask register timing signals are also indicated for masking during the read–write cycle. For further details of timing the reader should refer to the vendor specification for this product.

This 8M VRAM can do graphics at the rate of a 1.6GB/sec fill and 0.64GB/sec aligned Bit BLT. The timing for a 10-pixel vector draw is 4.1M vectors/second and for 7 × 8 character draw the rate is 4.5M characters/second.

The read/write bandwidth of the ×32 RAM with a 20 ns page cycle time is 200MB/sec. The "ultra fast" page mode on this part is similar to an output enable controlled Hyperpage (EDO) mode.

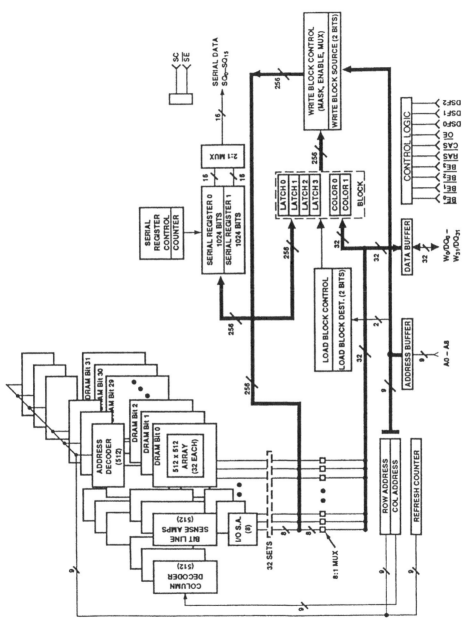

Figure 7.34   8M VRAM ("Window RAM") block diagram (source: Samsung [23])

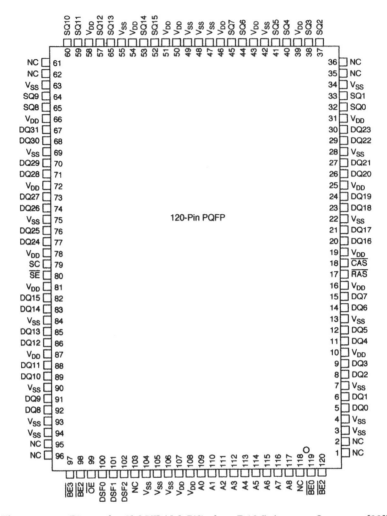

Figure 7.35 Pinout for 8M VRAM (Window RAM) (source: Samsung [23])

7.16 8M and 16M Synchronous VRAMs

An 8M synchronous VRAM has been standardized by JEDEC, but has not to this point been developed. At the present time any attempt at a multiport synchronous DRAM is more likely at the 16M density level. Such a part would be expected to continue the trend to faster RAM access and stripped-down features to reduce cost adders over the fast single port DRAMs.

IBM discussed a 64M synchronous VRAM. This part was constructed as eight 8Mb DRAM banks integrated with a gate array memory controller. It had a 64-bit parallel external bus to the host processor and a 32-bit serial external bus to the video display. The speed on the chip was 133 MHz and it used a 2.5 V power supply. It is not known if this part entered production.

$\overline{CAS}$ ⤵ *2	$\overline{RAS}$ ⤵ *1							Mnemonic Code	Function
$\overline{BE}_{3-0}$	$\overline{CAS}$	$\overline{OE}$	DSF2	DSF1	DSF0	RA_{8-0}	W_{31-0}		
X	0	X	X	X	0	X	X	RST	Reset Cycle
X	0	X	X	X	1	X	X	CBR	CBR Refresh
⇑ (Note 2)	1	0/1	X	X	0	Row	WPB Mask	RW	New Row Initiation for any RW Cycle
⇑ (Note 2)	1	0/1	X	X	1	Row	WPB Mask	TEST	Test Cycle

(a)

$\overline{BE}_{3-0}$ (Note 2)	DSF2	DSF1	DSF0	$\overline{CAS}$ ⤵ * 3 CA (8 7 6 5 4 3 2 1 0)	$W_{31}/DQ_{31}.$ W_0/DQ_0	Mnemonic Code	Function
⇑	1	0	0	X X X X X X X 0/1 0	Pixel Color Data	RCR	Read Color Reg 0 or 1
⇑	1	0	1	X X X X X X X 0/1 0	Pixel Color Data	LCR	Load Color Reg 0 or 1
⇑	1	0	0	X X X X X X X X 1	Mask Data	RMR	Read Mask Reg
⇑	1	0	1	X X X X X X X X 1	Mask Data (WPB)	LMR	Load Mask Reg
X	0	0	0	|←Column Address→| X 0 0	X X X X	UFBR	DRAM to Latch0
				|←Column Address→| X 0 1	X X X X		DRAM to Latch1
				|←Column Address→| X 1 0	X X X X		DRAM to Latch2
				|←Column Address→| X 1 1	X X X X		DRAM to Latch3
⇑	0	1	1	|←Column Address→| X 0 0		UFBWL	Latch0 to DRAM
				|←Column Address→| X 0 1			Latch1 to DRAM
				|←Column Address→| X 1 0	← Byte Mask →		Latch2 to DRAM
				|←Column Address→| X 1 1			Latch3 to DRAM
⇑	0	0	1	|←Column Address→| X 0 0	← Byte Mask →	UFBW8	From Color Reg 0 to DRAM
				|←Column Address→| X 0 1	← Byte Mask →		From Color Reg 1 to DRAM
				|←Column Address→| X 1 0	←Col Reg Select →		C0 = Fgrd, C1 = Bkgrd to DRAM
				|←Column Address→| X 1 1	←Col Reg Select →		C0 = Bkgrd, C1 = Fgrd to DRAM
X	0	1	0	|←Column Address→| X X 0	X X X X	SRT	Split Read Transfer
X	0	1	0	|←Column Address→| X X 1	X X X X	SRTR	Split Read Transfer with SAM Pointer Reset
⇑	1	1	0	← Column Address →	$D_{OUT}(Q_{31-0})$	UFR	Ultra Fast Page Read Cycle
⇑	1	1	1	← Column Address →	$D_{IN}(D_{31-0})$	UFW	Ultra Fast Page Write Cycle

⇑ = Byte Control   (b)

Figure 7.36 Simplified functional truth tables for 8M VRAM: (a) RAS\ control; (b) CAS\ control (source: adapted from Samsung [23])

7.17 Triple Port VRAM

A triple port RAM has been offered by Micron [13]. While the part is no longer actively being marketed, it is still in use with a small user base and will be mentioned in passing. This 1M part has a 256K×4 DRAM with 20 ns fast page mode access, and two 512 × 4 bidirectional SAMs. Data can be accessed at the RAM port, at either of the SAM ports, and transferred bidirectionally between the DRAM and either SAM. All three ports can be operated independently except during a transfer of data between ports.

The bidirectional aspect of the RAM and SAM have made this part of interest in data communications and networking applications as well as for high end graphics processing.

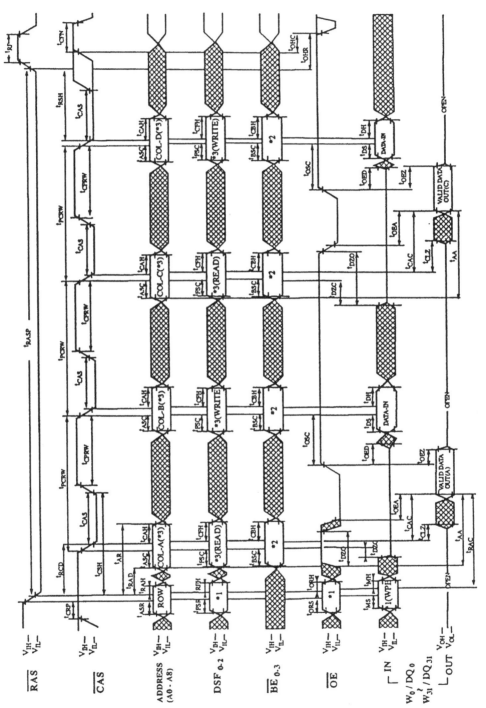

Figure 7.37 Output enable controlled (EDO) read–write cycle (source: Samsung [23])

7.18 VRAMs with Z-buffers

A few graphics RAMs have been developed for three-dimensional graphics applications. 3D computer graphics has been used primarily in engineering workstations, but is now becoming more popular in PCs for games and virtual reality applications. Two DRAM chips that are targeted at this applications area are described in this section.

7.18.1 3D-RAM

A 3D frame buffer DRAM with very fast rendering to 400 million pixels per second has been developed by Mitsubishi [25] primarily for the engineering workstation environment.

Standard DRAMs or VRAMs have been less than adequate in this application due to the additional bandwidth needed to process the large amount of data required for color and depth. Assuming a graphics workstation has 1280 × 1024 pixel screen resolution plus 24 bits per pixel for color (plus eight for color blend) and 32 bits for depth per pixel, the total bit requirement for a single frame is 10.2MB of memory.

The normal memory operation in 3D pixel rendering is the read–modify–write. This is because the old Z value must first be read. The old and new Z value are compared and the result written back to the memory. Comparison, for example, could entail deciding which data, old or new, is in front in the Z-direction since this is the information which must remain visible.

The 3D RAM combines on one chip a VRAM subsystem consisting of a four-bank DRAM and two SAM registers, a triple ported SRAM cache subsystem, a blend logic unit, and a compare logic unit. The SAM outputs refresh the screen. The cache subsystem consists of a tag cache, a data cache, and a compare unit.

New data moves from the processor into the compare and blend units. The new data is blended with the data in the cache, the blend is compared to the old data, and the result is transferred to the SRAM cache and thence to the DRAM banks. From the DRAM banks, the new frame is transferred to the SAM register and out to the video display. A 4 × 8 × 8 block transfer is available on the 3D RAM, but it is rare to change all 256 bits in the block in one block write.

A schematic block diagram showing the the 3D RAM internal data path in a "mostly write" situation is shown in Figure 7.38.

The 3D RAM handles data at a rate of 400M-pixels/sec.

3D RAM configurations for PC applications have appeared in production in high end graphics boards.

7.19 DDR SGRAMs

SGRAMs also exist as Double Data Rate (DDR) parts. The DDR SGRAM has the same 2-bit prefetch scheme on the output circuitry as the DDR SDRAM, which permits the burst datarate to be double the clock rate. The width of the databus is ×32 as shown in the pin configuration of a 64M DDR SDRAM in Figure 7.39.

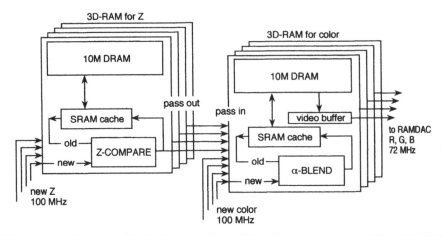

Figure 7.38 Schematic block diagram of 3-D RAM data path (source: Mitsubishi [26])

As discussed for the SGRAM I, the column block write is a feature which is present in the SGRAM and not in the SDRAM. A timing diagram for load color register and column block write for the DDR SFRAM is shown in Figure 7.40.

A 16-column block write is performed by issuing a block write command. On the next two cycles the column address mask is provided on the 32 DQ pins. The datastrobe is used to clock the DQ pins as shown. After the first block write command a NOP cycle is required, then on the following cycle another block write command or a normal write can be issued. Since this is a write operation, it is independent of the CAS\ latency.

The 32 bits of data written to the array is stored in the color register. This data is masked by the column address mask to provide the final data that is to stored. The color register must be loaded before the column block cycle begins.

The data mask function is used in conjunction with the normal write cycle. The data mask is stored in the mask register and is activated during a write cycle by taking the DM high. It masks the write data bits. This is shown in the timing diagram in Figure 7.41.

As shown in the figure, the data mask can be referenced on both the rising and falling edges of the datastrobe.

7.20 Dual Port Graphics DRAM

A dual port RAM has two independent ports with identical functionality. Mitsubishi developed a 16M Dual Port Synchronous Graphics DRAM targeted at the high performance graphics market. This part has an integrated cache SSRAM on chip with a 16M SDRAM with four independent banks.

Two identical 16-bit wide parallel ports can access the fast cache SRAM at the same time, as long as the same bit is not being read and written to. Byte masking is available with the DQM pins. An internal 32-bit write-per-bit mask and a block write mask are included. A block diagram of the dual Port Graphics DRAM is shown in Figure 7.42.

512Kx32x4Banks DDR

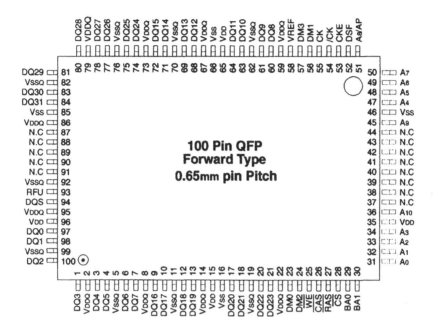

Product	64Mb DDR SGRAM		
	512K x 32 x 4Banks DDR		
# of Banks	4		
Bank Address	BA0, BA1		
Auto Prechage	A8/AP		
Row Address	A0~A10		
Column Address	A0~A7		
Packages	MS-002GC-2 / MO-212/A		

Figure 7.39 Pin configuration for a 64M 4 bank DDR SGRAM in 100 pin QFP package (source: JEDEC Standard 21C)

A 256-bit wide internal bus makes transfers from the SRAM to the four 4M banks of the DRAM. At 143 MHz, the effective bandwidth with two parallel 16-bit ports is 4.58GB/sec.

The part has a truth table for DRAM operations and a truth table for SRAM operations. The DRAM banks are not externally accessible. The DRAM truth table functions involve transfers to and from the SRAM and also refresh and mode register control.

The SRAM is externally accessible and could be read and written.

COLUMN BLOCK WRITE FOR DDR SGRAM

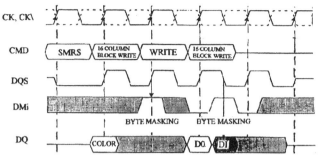

Figure 7.40   Column block write timing with column mask for DDR SGRAM (adapted from JEDEC Standard 21C)

7.21   Integrated Frame Buffers

Another direction that the DRAM frame buffer can go is to fuller integration of the DRAM with the graphics controller and the RAMDAC. One such circuit, from Neomagic, Inc. has been developed [27].

This chip combines a super VGA (SVGA) graphics accelerator, RAMDAC, frequency synthesizer and local bus interface (PCI, VL bus) with an 8Mbit DRAM. The chip takes advantage of the wide internal bus of the DRAM by having a 128-bit datapath between the DRAM and the controller. A later version had 20Mbit (2.5MB) of DRAM on the chip.

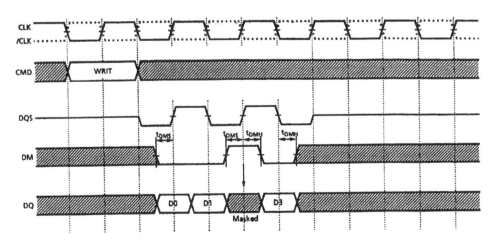

Figure 7.41   Data mask for the DDR SDRAM and SGRAM (source: Toshiba 64M DDR SDRAM))

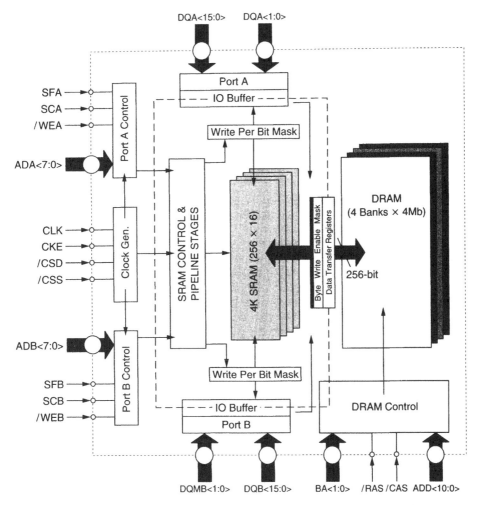

Figure 7.42 Dual Port Synchronous Graphics DRAM (source: Mitsubishi Dual Port SGRAM Datasheet)

Even at a comparatively slow speed, the bandwidth of such a chip can be significant. For example, running at 30 MHz on a 128-bit bus, such a chip would have a datarate of 500MB/sec. This is double the rate of a 66 MHz ×32 SGRAM running at 66 MHz (264MB/sec). The slower speed can also reduce power consumption, making this type of chip satisfactory for portable PC applications.

This integrated frame buffer was later upgraded to an integrated 2.5 MB DRAM array and further features were added.

Another integrated frame buffer for notebook computers was developed by Trident. The chip included a video accelerator, a 2D/3D graphics engine, a VGA controller, a display interface, and 2.5MB of SDRAM.

Other integrated graphics DRAMs were configured as massive parallel processors. An example is a chip from Accelerix which is configured as a SIMD (single input multiple data) processor. It takes advantage of the inherent parallel architecture of the DRAM array by associating each of 4096 pixel processors with 4 columns of a 4MB DRAM array.

Bibliography

1. JEDEC Standard 21C.
2. Toshiba, *Application Specific DRAM Data Book*, 1994.
3. Frame buffer architecture, *NEC Memory Products Databook, Volume 1, DRAMs, DRAM Modules, Video RAMs*, 1993.
4. 256K×4 Field Memory, TMS4C1060-30, *Texas Instruments MOS Memory Databook*, 1993.
5. Micron Technology, *8M SGRAM preliminary datasheet*.
6. Toshiba, *8M SGRAM preliminary datasheet*.
7. Toshiba, *Specialty DRAM Databook*.
8. United Memories, *4M SGRAM Preliminary Data Sheet*.
9. NEC (uPD481850), *8M SGRAM Preliminary Datasheet*.
10. H. Kotani, *et al.*, A 50 MHz 8-Mbit video RAM with a column direction drive sense amplifier, *Journal of Solid State Circuits*, Vol. 25, No. 1, February 1990, 30–35.
11. H. Kotani, *et al.*, A 256-Mb DRAM with 100 MHz serial I/O ports for storage of moving pictures, *Journal of Solid State Circuits*, Vol. 29, No. 11, November 1994, 1310.
12. *NEC Memory Products Data Book, Volume 1 2, DRAMS, DRAM Modules, Video RAMs*, 1993.
13. Micron Technology, *Specialty DRAM Graphics/Communications Data Book*, 1993.
14. Samsung, *DRAM Databook*, 1995.
15. Hitachi, *DRAM Databook*, 1993.
16. Prince, B., *Semiconductor Memories*, 2nd Edition, John Wiley & Sons, 1992.
17. United Memories, *4M Synchronous Graphics DRAM Preliminary Datasheet*, 1994.
18. Texas Instruments, *MOS Memory Databook*, 1995.
19. R. Myravaagnes, Rambus memories shipping from Toshiba, *Electronics Products*, November, 1993
20. R. Wilson, Cirrus unwraps Rambus graphics line, *Electronic Engineering Times*, 26 June, 1995, 16.
21. IBM Microelectronics, *4Mb VRAM Data Sheet and Applications Notes*, 1994
22. Texas Instruments, *4-Megabit Video RAM, Application Guide*, 1993.
23. Samsung Semiconductor, *256K × 32 CMOS Window RAM Preliminary Datasheet*, September 1993.
24. Micron Technology, *1995 DRAM Data Book*.
25. K. Inoue, *et al.*, A 10Mb 3D frame buffer memory with Z-compare and alpha-blend units, *IEEE International Solid-State Circuits Conference*, 1995, 302.
26. T. Watanabe, *et al.*, 3-D CG media chip: an experimental single-chip architecture for three-dimensional computer graphics, *IEICE Trans. Electron.*, VOL. E&&C, No. 12, December 1994, 1881.
27. Calle R., Graphics IC for portable PCs brings frame buffer on chip, *Electronics Products*, May 1995, 103.
28. K. Kushiyama, *et al.*, A 500 Megabyte/s Data-Rate 4.5M DRAM, *IEEE Journal of Solid State Circuits*, Vol. 28, No.4, April 1993, 409.
29. Samsung, 128 × 32bit × 2 Bank Synchronous Graphics RAM preliminary data sheet.
30. Mitsubishi Semiconductor, *16M SGRAM Datasheet*, 1997.
31. MoSys Inc., *16M SGRAM Datasheet*, 1998.
32. IBM, *DDR SGRAM Datasheet*, December, 1997.

33. R. Torrance, *et al.*, A 33 GB/s 13.4M-bit integrated graphics accelerator and frame buffer, *IEEE International Solid-State Circuits Conference Proceedings*, 1998.
34. Mitsubishi Semiconductor, *Dual-Port DRAM Datasheet*, 1997.
35. W.K. Luk, *et al.*, Development of a high bandwidth merged logic/DRAM multimedia chip, *IEEE Conference on Computer Design*, October, 1997, 279.
36. *JEDEC JESD21C for DDR SGRAM.*
37 *Toshiba 64M DDR SDRAM.*

8 Power Supply, Interface, and Test Issues

Along with the rapid increase in speed of components in the average computer system have come both the lower power supply voltages needed to reduce power dissipation and a requirement for new high speed interfaces to reduce the transmission line effects in fast systems. While none of the high speed interfaces available appears yet to have entered in the mainstream, several have appeared on fast memories.

8.1 Different Voltages in the System

Power supply voltages today are a mix of 5 V and 3.3 V in the average computer. It appears that it will be the end of the decade before appreciable numbers of systems using high speed processors and memories convert to 2.5 V or lower. Figure 8.1 shows an estimate for the trends in voltages for commodity RAMs.

A major driver of the trend toward 3.3 V is the PC systems with 3.3 V processors. As a result, some memory products, such as the 3.3 V burst synchronous SRAMs and 3.3 V 16M DRAMs that are used with this processor, are leading the change. 3.3 V 16M DRAMs are also used in the 686 class processor systems.

While consensus of the PC systems designers seems to be that both voltages will need to be present in most PC systems for at least the next generation to offer the widest range of choices for all components in these commodity systems, the cache SRAMs, DRAMs and processors are being shipped at the lower voltage.

The problem is that all 3.3 V parts needed for a single system are in many cases either not available or not cost equivalent with the 5 V parts. Since the cost of mixing parts is that of a voltage converter logic chip, all 3.3 V parts will be used when

sum cost 3.3V chips < sum cost of mixed voltage chips + converter

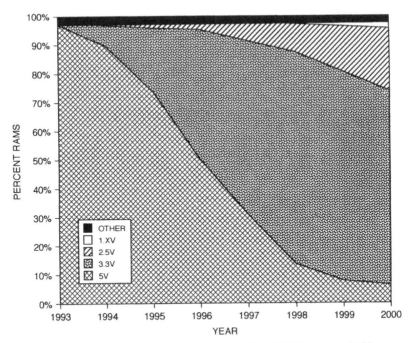

Figure 8.1   Voltage trends for commodity RAMS (source: MSI)

Meanwhile one solution for the continued requirement for 5 V tolerant systems has been to make RAMs with 3.3 V power supply and LVTTL interface whose I/O will function with parts having 5 V power supply and TTL interface.

As the processors move to even lower voltages in the next few years, the PCs are likely to continue to lead the way to lower voltages. Chips such as the burst SSRAM, which are used primarily in PCs can give an indication of the overall trend. A projection of the 1M burst SSRAM usage split by voltage is shown in Figure 8.2.

Let us compare some of the parameters for a 5 V and a 3.3 V 64K×18 synchronous cache SRAM in the same technology to see why the move to 3.3 V was made.

Comparing the power dissipation, the power supply current (ICC) for the 5 V part at 12 ns cycle time is 215 mA which is 1.1 W power dissipation. The power supply current for the 3.3 V part at maximum cycle (12 ns) is 180 mA which is 0.6 W. The active power dissipation of the 5 V part is roughly twice that of the 3.3 V part running at the same speed. One reason that the power supply voltage was dropped for fast parts was to reduce the active power dissipation.

A drawback is that dropping the voltage slows the part down if the same technology is used for both. The maximum clock to output cycle time of the 5 V part is 8.5 ns with no output load, and that of the 3.3 V part is 12.5 ns with no output load.

In the next level of technology, however, the 3.3 V part will regain the speed yet still dissipate only half the power. So, for example, the power dissipation of a 5 V SRAM running with an 8.5 ns cycle is about 1.5 W but that of the 3.3 V SRAM running at the same 8.5 ns cycle time would be around 0.8 W.

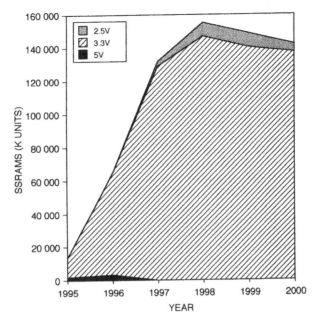

Figure 8.2   1M burst SSRAM shipments split by voltage (thousand units) source: Memory Strategies International)

The gain in moving to the tighter technologies is in speed and the gain in going to 3.3 V at these speeds is in active power dissipation.

8.2   Fast Interfaces

8.2.1   *Established Interfaces*

The current most commonly used interfaces are shown in Table 8.1. They include the 5 V TTL, 5 V CMOS, 3.3 V LVTTL and 3.3 V LVCMOS; 2.5 V and 1.8 V interfaces are also standardized but not yet in common usage. For fast systems which can use a

Table 8.1   Common electrical interfaces (V)
(DC operating characteristics)

	TTL	CMOS	LVTTL	LVCMOS
VDD (typ.)	5	5	3.3	3.3
VOH (min.)	2.4	VDD−0.2	2.4	VDD−0.1
VOL (max.)	0.4	0.2	0.4	0.1
VIH (min.)	2.0	0.7VDD	2.0	2.0
VIL (max.)	0.8	0.2VDD	0.8	0.8

bipolar technology, there is also ECL (Emitter Coupled Logic) and BTL (Backplane Transceiver Logic). Since memories using BTL are rare, this interface will only be noted in passing.

Minimum output voltage swings are 4.1–5.1 V for CMOS, 2.0 V for TTL, 2.8–3.4 V for LVCMOS, and 2.0 V for LVTTL.

Maximum output voltage swings are VDD–2VBE for TTL, where VBE = 0.7 V at 25°C ambient, VDD–0.2 for CMOS, VDD–0.1 for LVCMOS, and VDD–VT(n) for LVTTL where VT(n) = 0.7 V.

The maximum VOH of TTL compatible CMOS circuits is either VCC–2VT(n) or VCC–VT(n). The value of VOH(max.) is used for calculating voltage reflections due to transmission link effects and voltage coupling (noise) due to crosstalk.

A full rail-to-rail swing for a high speed RAM consumes too much power regardless of whether it is driving lumped capacitance loads or terminated lines. For example, driving rail-to-rail with 3.3 V supplies and 100 pF lumped loads at 100 MHz requires the dissipation of over 100 mW per output driver.

Since memories can have from four to 32 output drivers operating simultaneously, the power consumption could be up to 3 W at 100 MHz.

A typical capacitive load can also result in the RAM being significantly slower than is specified in a datasheet for conditions of no output load. Characteristic curves published in some datasheets can give an idea of the reduction in access time with capacitance. Figure 8.3a illustrates a specified increase in access time for a fast 32Kx8 SRAM with change in load capacitance.

RAMs in fast systems can also have transmission line problems. The output transition times for 100 MHz are usually around 2 ns and times of flight down a 15 cm wire are typically 1 ns. This causes reflection problems affecting signal integrity if termination is not used.

Figure 8.3b(i) illustrates a typical lumped load transmission line representation. Figure 8.3b(ii) shows a derating curve for change in output rise and fall times for a lumped capacitive load [8].

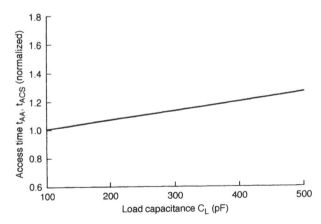

Figure 8.3a Characteristic curve showing SRAM access time vs. loading (source: Hitachi)

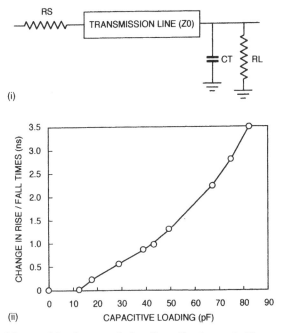

Figure 8.3b Typical lumped load transmission line: (i) schematic illustration; (ii) output rise and fall times for fast SRAM with lumped capacitive load (source: Motorola)

In Figure 8.3b(i) Z_0 is the characteristic impedance of the transmission line, R_L is the line termination, and CT is the transmission line capacitance.

The three conditions are:

1. If $R_L = Z_0$ the line is critically damped and there is no reflection.
2. If $R_L > Z_0$ it is underdamped and there is a reflection.
3. If $R_L < Z_0$ it is over damped and there is no reflection.

In the second case the line should be terminated.

A series of point loads can be represented as a lumped load where Z (square root of L/C') is an electrically equivalent impedance to the point loads on an actual line as transformed by the line characteristics, where

$$C' = \text{sum } (C_1 = C_2 + \ldots + C_t)$$

The power dissipation is given by

$$\text{Power} = \tfrac{1}{2}CV^2f \qquad f = \text{frequency}$$

Since power is a factor of the square of the voltage, the power problem in high speed systems can be helped by limiting the swing of the RAM interface.

Returning to our consideration of high speed interfaces, for speeds up to 20 MHz, TTL was a "low swing" interface compared to CMOS for a 5 V power supply part.

At 3.3 V power supplies, however, the 2 V swing of LVTTL is 60 percent of the LVCMOS power supply level for a 3.3 V part. With careful attention to system design, however, LVTTL can be used up to about 100 MHz.

An example of the ac characteristics and output loading of a fast, high density DRAM using the 3.3 V LVTTL interface is shown in Figure 8.4. Figure 8.4(a) shows the set-up and hold characteristics of a typical 256M SDRAM I. The reference cross-over point is 1.4 V. The access time from clock (t_{AC}) and oautput hold time from clock (t_{OH}) are shown.

Figure 8.4(b) shows a typical terminated and unterminated output loading circuit schematic for this LVTTL intnerface SDRAM. For the terminated load, ac timing tests have VIL = 0.4 V and VIH = 2.4 V, with the timing referenced to the 1.4 V crossover point. Ac measurements assume that the transition time (Tt) = 1.0 ns. For the unter-minated load, ac measurements assume Tt = 1.2 ns.

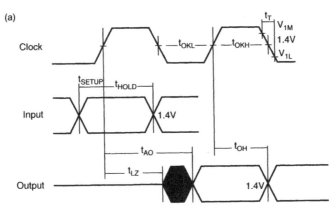

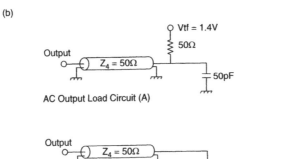

Figure 8.4   Ac characteristics and output loading for a typical LVTTL interface: (a) set-up and hold characteristics; (b) top: terminated load; (bottom: unterminated load (source: IBM 256M SDRAM Datasheet)

For higher system speeds, the usual choice for high speed interfaces has been ECL. Its limited swing and ability to drive terminated lines made it superior in high speed systems to the TTL interface. ECL works well even with small 800 mV output swing since the differential inputs have excellent common mode voltage rejection characteristics. For high speed systems, that can afford the cost of the memory, ECL memory with one of the common ECL interfaces can be used. There are older 100K or 10K (101K) 5 V generation interfaces and a newer 300K ECL 3 V generation interface. These interfaces are shown in Table 8.2.

ECL and BTL are both intended for bipolar circuit technology. While BTL memories are rare, there are small, very fast bipolar RAMs which use the various ECL interfaces. The pinouts of a 10K ECL and a 100K ECL bipolar ECL SRAM are shown in Figure 8.5(a) and (b). A typical output test loading diagram is shown in Figure 8.5(c) [5].

Notice that the VCCA power supply is ground, R_1 is 50 ohms, and C_1 is 30 pF. VCCA is for the array and VEE is for the outputs. Figure 8.5(a) (101K ECL) shows a standard ECL type pinout with power and ground pins on the ends of the package. Figure 8.5(b) shows a pinout with the power and ground pins at the center of the package. Both parts are specified in the 5–7ns access time range.

Bipolar ECL interfaces can also be added to CMOS RAMs. However, the increased cost of adding Bipolar transistors to CMOS technology kept the industry searching for high speed CMOS interfaces.

8.2.2   *Newer High Speed Interfaces*

A number of newer interfaces that use CMOS circuitry can give the ECL advantage of low swing and ability to drive terminated lines. All of these interfaces have a

Table 8.2   Standard interfaces for bipolar integrated circuits (V)
(all ECL values are negative)

	10 (101)K ECL		100K ECL	300K ECL	BTL
Power supply voltage:					
VDD	5.2		4.5	3.5	5
Interface characteristics:					
Output swing	0.8		0.8	0.8	1.0
NMH	0.1		0.1	0.1	0.5
NML	0.15		0.15	0.15	0.4
DC operating conditions:					
	All ECL		BTL		
VOH	0.88	2.1			
VOL	1.81	1.62	1.1		
VIH	1.165	0.88	1.62		
VIL	1.81	1.475	1.47		
VT	1.38	1.26	1.55		

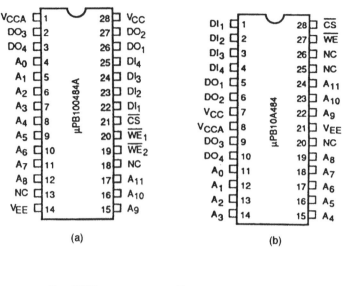

(a)   (b)

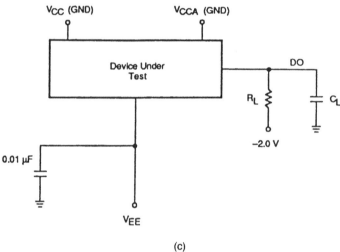

(c)

Figure 8.5   Pinouts for (a) 101K and (b) 100K ECL SRAMs (source: NEC), and (c) typical output test loading diagram

restricted signal swing of the order of 1 V or less; however, the actual inputs and output logic levels vary widely.

In CMOS technology the JEDEC standard low swing interfaces are GTL (Gunning Transceiver Logic), CTT (Center Tap Terminated Logic), and HSTL (High Speed Transceiver Logic). Table 8.3 shows the DC operating characteristics of these interfaces. LVTTL and BTL are also shown for comparison, along with the newer 2.5 V normal interface shown for an output drive current of $+/-$ 1 mA.

The output swing for 3.3 V CTT and for 3.3 V GTL is 0.8 V which is less than half of the 2.0 V output voltage swing for 3.3 V LVTTL. GTL, which was developed by

Table 8.3 Standard high speed bus interfaces for memory (DC operating characteristics)

	2.5 V LVTTL	3.3 VLVTTL	CTT	BTL	GTL	HSTL
VOH (min)	2.0	2.4	1.9	2.1	1.2	1.1
VIH (min)	1.7	2	1.7	0.85	0.85	0.85
VREF	1.2	1.5	1.5	1.55	0.8	0.75
VIL (max)	0.7	0.8	1.3	1.47	0.75	0.65
VOL (max)	0.4	0.4	1.1	1.1	0.4	0.4
VDD (Q)	2.5	3.3	3.3	5	1.2	1.5
NM(H)	0.3	0.4	0.2	0.48	0.35	0.25
NM(L)	0.3	0.4	0.2	0.37	0.35	0.25
SWING(O)	1.6	2	0.8	1	0.8	0.7

Xerox, is also applicable at a 2.5 V power supply level. Single ended HSTL is also shown.

The input and output levels of these interfaces are plotted in Figure 8.6 for easier comparison.

The termination (reference) voltage for CTT is 1.5 V which is close to the center voltage for LVTTL. CTT, developed by Mosaid, can therefore be considered a subset of LVTTL with tighter input and output swings and smaller noise margin.

BTL, or Backplane Transceiver Logic, is used in systems conforming to the IEEE Futurebus standard which is a standard for bipolar logic. This is not very applicable to memories because of the high cost of bipolar memories.

GTL is more like a level shifted (compressed) BTL. The noise margins of GTL at 350 mV are slightly narrower than the noise margins for BTL (480 mV); however, GTL could be considered a progression of BTL. HSTL is similar to GTL in output swing and reference level, but has widened the narrow GTL input swing.

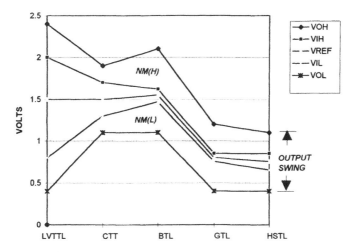

Figure 8.6 Plot of high speed bus interface levels

Circuit diagrams of possible output drivers for CTT and GTL are shown in Figure 8.7.

The CTT driver shown in Figure 8.7(a) can handle both standard and reduced signal levels. The termination voltage, VTT, which controls the levels of the logic swing, can be as high as 5 V.

In the GTL interface circuit shown in Figure 8.7(b), when the NMOS driver is turned off, the resistor pulls the node up to the nominal 1.2 V level of VTT. The receiver then differentially compares an incoming signal to V_{REF} which is typically 0.8 V.

Another differential interface that has been standardized is the HSTL differential interface. The fast SRAMs were the first high speed memories to use the HSTL interface.

The HSTL standard (JESD8-6) is intended as a guide for accurate testing of the high speed memory devices to permit replicability of data. It is specified for both DC and AC test conditions and is specified for several classes of output buffers. These include the following [7]:

Class 1:   push–pull output buffers for unterminated loads, and for symmetrically parallel terminated loads

Class II: push–pull output buffers for source series terminated loads, and for symmetrically double parallel terminated output loads

Class III: push–pull output buffers for asymmetrically parallel terminated loads and for asymmetriacally parallel terminated output loads

Class IV: push–pull output buffers for asymmetrically double parallel terminated loads

HSTL supply and I/O levels are shown in Figure 8.8(a) and differential input levels are shown in Figure 8.8(b).

The first fast DRAMs to appear with a low swing interface were synchronous DRAMs that used the CTT (also called T-LVTTL) interface [10].

Pinouts for the topside view of a 16M (4M×4) SDRAM and an 8M (256K×32) SGRAM using the T-LVTTL (CTT) interface are shown in Figure 8.9.

Note that the pinouts reflect the speed function of the part. Multiple power and ground pins are added. Power and ground for the outputs (V_{DDQ} and V_{SSQ}) either alternate with outputs or are every two outputs. Power and ground for the array are separate (V_{DD},V_{SS}) on the address side of the package as well as the opposite side of the package. Control and clock pins are clustered in the center of the packages and spaced at least a pin away from the address pins.

The speed (output valid from clock) that is specified for the Fujitsu 4M×4 SDRAM is 7 ns for LVTTL and 6 ns for T-LVTTL, indicating a 17 percent speed improvement for the part with the terminated interface.

The LVTTL and T-LVTTL interface values can be seen in the recommended operating conditions for the Fujitsu 16M SDRAM shown in Table 8.4. All values are referenced to ground (V_{SS}).

An output load for the unterminated LVTTL interface is shown in Figure 8.10(a) and a terminated lumped capacitance load for the T-LVTTL SDRAM is shown in Figure 8.10(b). A terminated point-to-point bus design for the T-LVTTL SDRAM is shown in Figure 8.10(c). The termination resistor is specified at 50 ohm.

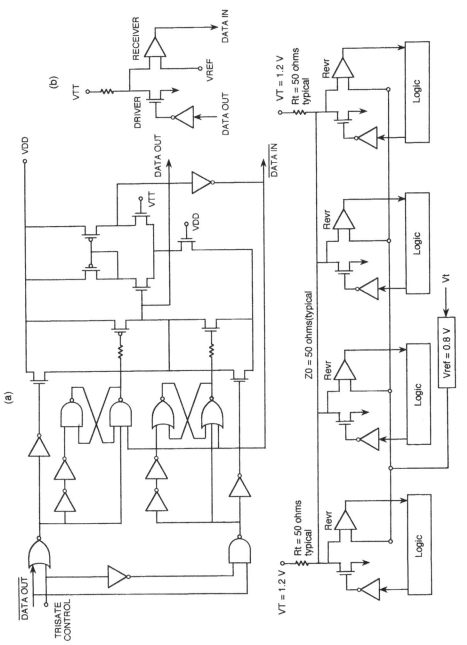

Figure 8.7 Output drivers for (a) CTT, and (b) GTL, and (c) multiple GTL tranceivers (source: Cypress [9])

Table 8.4   T-LVTTL (CTT) recommended operating conditions

Parameter	Symbol	Min.	Typ.	Max.	Unit
Supply voltage	V_{CC}, V_{CCQ}	3.0	3.3	3.6	V
	V_{SS}, V_{SSQ}	0	0	0	V
Reference voltage	V_{REF}	1.35	1.5	1.65	V
Terminated LVTTL input high voltage	V_{IH}	V_{REF} +0.2	–	V_{CC} +0.2	V
Terminated LVTTL input low voltage	V_{IL}	−0.3	–	V_{REF} −0.2	V
LVTTL input high voltage	V_{IH}	2.0	–	V_{CC} +0.2	V
LVTTL input low voltage	V_{IL}	−0.3	–	0.8	V
Termination resistor	R_T	–	50	–	O
Ambient temperature	T_A	0	–	70	8C

Source: Fujitsu [10]

Advance datasheets of SDRAMs with a GTL interface were also issued, but it is not certain that parts actually appeared on the market. Logic chips with the GTL interface, however, have been in production for some time. The pinout is the same as the standard pinouts shown above.

We have thus far considered the speed of fast memories outside the system. It is their speed in the system which counts, of course, and that can differ from the

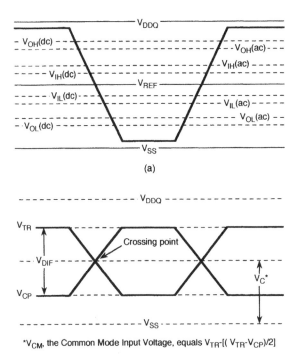

*V_{CM}, the Common Mode Input Voltage, equals V_{TR}-[(V_{TR}-V_{CP})/2]

Figure 8.8   (a) HSTL supply and I/O levels; (b) HSTL differential input levels (source: EIA JEDEC Standard 8 [7])

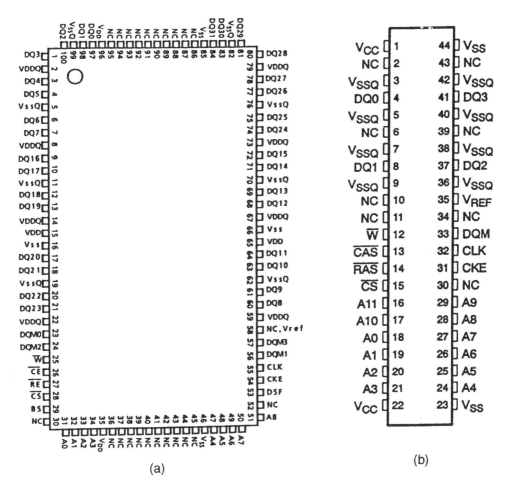

Figure 8.9 Pinouts of two SDRAMs with CTT or GTL interface (a) 256K×32 SGRAM; (b) 4M×4 SDRAM (source: Fujitsu [10])

specified speed for a number of reasons including transmission line effects in the system, chip skew, clock skew, and actual output loading in the system.

A high rate of data transmission requires matching the DRAM bandwidth to that of the processor. The logic interfaces between the two match and care must be taken in the design of the board.

Processors are in production with speed capabilities over 350MHz now. The fastest of these are expected to approach 800MHz by the year 2000. In cases to date where the speed on the memory bus matches that of the processor, either the cache has been integrated onto the processor chip as in the case of L1 cache, or a private bus has been used. For example, the Intel P6 processor uses a two-chip package that mounts both the processor and the cache SRAM chip together in a common package. A high speed interface (a variant of GTL) is used between the two. It is less common for separate components with high speed interfaces to be

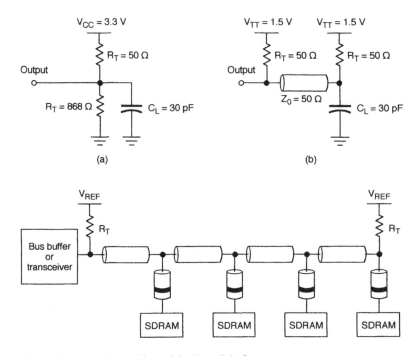

Figure 8.10 Output loads and bus design for LVTTL and T-LVTTL interface: (a) output load for unterminated LVTTL; (b) terminated lumped capacitive load for T-LVTTL; (c) bus design for T-LVTTL (source: Fujitsu [10])

used in a commercial system except in the case of very high end systems, super-computers and mainframes.

Stub Series Terminated Logic interfaces (SSTL) are defined for modules containing multiple high speed SDRAMs on a parallel bus. The presence of stubs on the module presents a different transmission line problem than do the "point-to-point" connections used commonly between the processor and SRAM cache.

The SSTL interface definition includes criteria for both dc and ac logic levels. The ac values are chosen to indicate the levels at which the receiver must meet its timing specifications. The dc values are chosen at a looser limit such that the final logic state is unambiguously defined.

A waveform illustrating the relationship between the various logic switching levels for SSTL is shown in Figure 8.11.

Once the receiver input has crossed the dc limit value, the receiver will change to and maintain the new logic state, even though there may be some ringing for a time about the tighter ac limit. This way, it can be fairly certain that the device will switch state a given amount of time after the input has crossed the ac threshold and not switch back as long as the input stays beyond the dc limits.

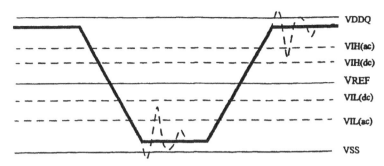

Figure 8.11 Logic switching levels for SSTL (source: EIA JEDEC Standard 8-8)

SSTL3 is the 3.3 V based high swing interface commonly used as a standard interface on the SDRAMs running faster than 133 MHz. Fast SDRAMs of 16M, 64M, and 256M densities are expected to use this interface.

V_{DDQ} for SSTL3 is defined as a 3.3 V power supply. That is, V_{DDQ} maximum is 3.6 V, V_{DDQ} minimum is 3.0 V, with typical at 3.3 V. The input dc and ac voltage levels are shown in Table 8.5.

SSTL2 is a 2.5 V based high speed interface which is expected to be used primarily on 64M and 256M SDRAMs. V_{DDQ} maximum is 2.7 V and V_{DDQ} minimum is 2.3 V. Other associated supply voltage levels are shown in Table 8.6.

This is configured in a similar way to the SSTL3 interface with input logic levels as shown in Table 8.7.

There are various classes of output buffers specified which depend on the specific circuit configuration. The JEDEC Standard 8 should be consulted for more details. This standard can be found on the JEDEC web page at < **www.jedec.org** >.

The power supply on the DRAM array may still be 3.3 V and the outputs can be SSTL2-compatible since there is no specific V_{DD} supply voltage requirement for SSTL2 compliance. It is required, however, that V_{DDQ} must be less than or equal to V_{DD} under all conditions.

The value of V_{REF} may be selected by the user to provide optimum noise margin in the system. V_{REF} is typically expected to be about 50% of the V_{DDQ} of the transmitting device and V_{REF} is expected to track variations in V_{DDQ}.

Table 8.5 Input logic levels for SSTL3

Symbol	Parameter	Min.	Max.	Units
Input dc logic levels				
V_{JH} (dc)	dc input logic high	$V_{REF} + 0.20$	$V_{DDQ} + 0.3$	V
V_{IL} (dc)	dc input logic low	0.30	V_{REF} 0.20	V
Input ac logic levels				
V_{IH} (ac)	ac input logic high	$V_{REF} + 0.40$		V
V_{IL} (ac)	ac input logic low		$V_{REF} -0.40$	V

Source: EIA JEDEC Standard 8-8.

Table 8.6 Supply voltage levels for 2.5 V (SSTL2)

Symbol	Parameter	Min.	Nom	Max.	Units	Notes
V_{DD}	Device supply voltage	V_{DDQ}		n/a	V	1
V_{DDQ}	Output supply voltage	2.3	2.5	2.7	V	1
V_{REF}	Input reference voltage	1.15	1.25	1.35	V	2, 3
V_{TT}	Termination voltage	$V_{REF}-0.04$	V_{REF}	$V_{REF} + 0.04$	V	4

Source: EIA JEDEC Standard 8-9.

Table 8.7 2.5 V SSTL2 input logic levels

Symbol	Parameter	Min.	Max.	Units
Input dc logic levels				
V_{IH} (dc)	dc input logic high	$V_{REF} + 0.18$	$V_{DDQ} + 0.3$	V
V_{IL} (dc)	dc input logic low	-0.30	$V_{REF}\ 0.18$	V
Input ac logic levels				
V_{IH} (ac)	ac input logic high	$V_{REF} + 0.35$		V
V_{IL} (ac)	ac input logic low		$V_{REF}\ -0.35$	V

Source: EIA JEDEC Standard 8-9.

8.2.3 Packet Protocol DRAM Interfaces

The two types of packet protocol DRAMs, the SLDRAM and the Rambus DRAM use differential interfaces.

The SLIO (Synchronous Link Input/Output) is used on the SLDRAM. The voltage levels of the SLIO are similar to the SSTL2 interface standardized by JEDEC for fast 2.5 V DRAMs. The implementation is slightly different due to variations in the I/O architecture of the SLDRAM and the JEDEC SDRAMs. The SLDRAM also includes the V_{OL} and V_{OH} correction capability of the part in the definition of the interface.

The Rambus DRAM also uses a low swing interface called Rambus Tranceiver Logic (RTL). The dc characteristics for the RTL interface are shown in Table 8.8.

Table 8.8 RTL interface dc characteristics

V_{DD}	Supply voltage	2.5 +/− 0.125V
V_{TERM}	Termination voltage	1.8 +/− 0.09 V
V_{REF}	Reference voltage	1.4 +/− 0.07 V
V_{IL}	Input low voltage	$V_{REF} - 0.5$ to $V_{REF} - 0.2$
V_{IH}	Input high voltage	$V_{REF} + 0.2$ to $V_{REF} + 0.5$

Source: Rambus Inc, RDRAM Datasheet.

8.3   Difficulties in Specification of High Speed Components

The setting of usable standards for high speed interfaces presents many practical difficulties. It is not sufficient to demonstrate that a particular component standard will result in working systems. Memory components, particularly DRAMs, are widely sourced commodity parts and must be testable in ways which allow them to be given worst case specifications that will guarantee their working under realistic system conditions in the same system with the components from other manufacturers also using that standard.

With restricted swing interfaces, test for worst case specification becomes difficult. Already components with the first low swing interfaces are beginning to present this type of test and specification problem.

8.4   Testing High Speed Memories

The problem of testing high speed memories can be broken into several parts. First of all, RAMs are getting faster externally at a higher rate than the speed gained internally in the RAM array. This means that a tester can test the internal functionality of the memory at a lower speed than it must test the interface circuitry. Core testers targeted at testing the functionality of the DRAM are being employed in many cases with a separate test for the high speed interface.

A second factor is that significant amounts of logic are being added on the high speed RAMs, such as DLLs, PLLs, registers, etc., which must also be tested. RAM and logic testers have evolved separately. Now, however, new testers that can check both RAM and logic are becoming available.

A third factor in memory testing is the introduction of entire subsystems integrated on the chip with the RAM. In this case, it is frequently difficult to isolate the memory and logic for separate external test and a test generator for the RAM must also be integrated on the chip.

A final problem is that as RAMs get faster, they are also getting denser. This presents a further problem in testing, as it takes longer to perform even standard tests on them.

8.4.1   Testing High Density Memories

As the RAMs get faster, they are also getting denser. This presents a further problem in testing the newer RAMs as it takes longer to perform even standard tests on the memories.

For example for a 16Kbit DRAM, it took 16384 accesses to obtain all of the data stored in the DRAM. At 400 ns cycle time, it took 6.5 ms to access the entire contents of the DRAM. However, for a 64K DRAM, the cycle time was only improved to about 270 ns, but it took 65536 accesses to access all the data in the DRAM or 17.7 ms which is almost three times as long.

Table 8.9 shows that by the 64M DRAM it will take 55 seconds to access the entire memory of one ×1 organized DRAM at the read cycle time of a DRAM. Even for a ×16 organized DRAM, it will still take 3.4 seconds. The problem is that the amount of the data which can be stored in a DRAM is quadrupling every three years on the average while the cycle time required for a single access is decreasing by only a small amount.

With modern processors requiring data at a much higher rate, it has been necessary to find a way to access more of the data in the increasingly larger DRAMs in a given time period cost-effectively.

The ability to test a high volume of DRAMs cost-effectively also depends on this effort.

8.4.2   Testing Stand-alone RAMs using BIST

Several examples have been shown of DRAMs and SRAMs which use internal test generators on the memory chip.

An example of a DRAM that uses a Built-In-Self-Test (BIST) generator was given by Texas Instruments in 1997 at the ITC. This 256M SDRAM used a BIST generator for lengthy test algorithms that would add significantly to cost on the external tester. Figure 8.12 illustrates the block diagram for this 256M SDRAM showing the BIST generator. The BIST generator was activated during burn-in. The advantage of this use of BIST is the reduction in test time and, hence, cost in terms of time spent on an expensive test machine. The disadvantage of this use of BIST is the inability to distinguish between burn-in failures and functional test failures.

8.4.3   Testing with Boundary Scan

One solution for testing RAMs with limited access to the array such as serial access memories is to use boundary scan techniques. Boundary scan involves adding four additional pins to the RAM. This can be done in specialized parts but not in components with standard pinouts such as the commodity DRAMs and SRAMs.

Table 8.9   Total time to access the data in a ×1 DRAM

Density (bits)	T_{CYC} (ns)	Time to access ×1 DRAM (ms)
16K	220	3.6
64K	180	11.8
256K	150	39.3
1M	130	136.2
4M	110	461.3
16M	95	1593.7
64M	85	55703.8

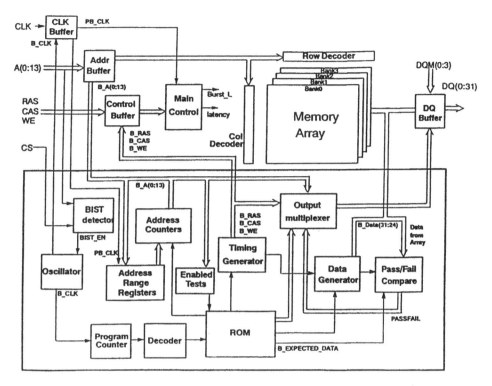

Figure 8.12   Block diagram for a 256M SDRAM showing the BIST generator (source: Texas Instruments)

An example of an SRAM that used boundary scan test is a synchronous 8K×8 line buffer. This part has the IEEE Standard 1149.1 Test Access Port shown in the block diagram in Figure 8.13(a) [6]. The four extra pins into the test access port are shown. They are the test mode select (TMS) pin, the test clock input (TCK), test data output (TDO) and Test data input (TDI). There are two classes of instruction. These are standard defined instructions and device specific instructions. Figure 8.13(b) illustrates the boundary scan paths.

8.4.4   *Testing High Speed RAMs*

In order to properly test ac performance in fast RAMs, the ac loading must be changed to a transmission line terminated with a characterisic impedance matching the load. Reflections are then cancelled and maximum noise margin is obtained.

The characteristic impedance of the terminated transmission line should match that of the source signal so that reflections are cancelled and maximum noise margin is obtained [8].

Ac test conditions for SSTL3 are specified as part of the JEDEC standard for the SSTL3 interface. The ac input test signal waveform is shown in Figure 8.14(a) and the ac input test condtions are shown in Figure 8.14(b).

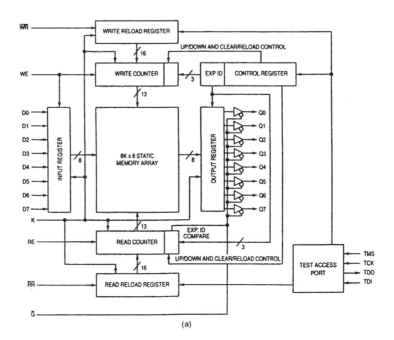

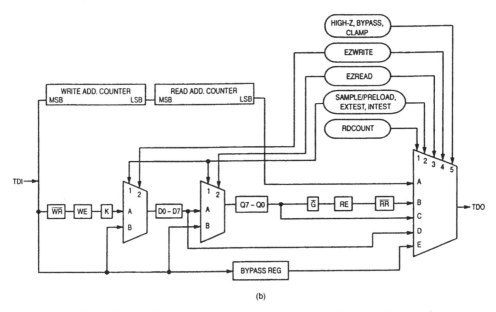

Figure 8.13   8K×8 line buffer using boundary scan: (a) block diagram; (b) boundary scan paths (source: Motorola [6])

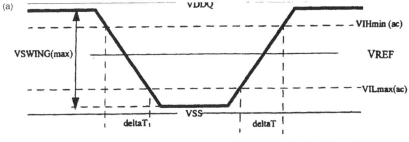

$$SLEW = (VIHmin(ac) - VILmax(ac)) / deltaT$$

(b)

Symbol	Condition	Value	Units	Notes
VREF	Input reference voltage	0.45 * VDDQ	V	1,4
VSWINGmax	Input signal maximum peak to peak swing	2.0	V	1,2
SLEW	Input signal minimum slew rate	1.0	V/ ns	3

Figure 8.14   AC test conditions for the SSTL3 interface: (a) ac input test signal waveform; (b) ac input test conditions (source: EIA JEDEC Standard 8-8)

In every case the input waveform timing is referenced to the input signal crossing through the reference voltage level which is applied to the device under test. The 1V/ ns input signal minimum slew rate must be maintained for the input signal swing in the range between $V_{ILmax}(ac)$ and $V_{IHmin}(ac)$.

It is understood that these ac test conditions may be measured under nominal voltage conditions, as long as analysis shows the RAM will meet its timing specifications under the specified voltage conditions.

Ac test conditions for the SSTL2 interface are given in Table 8.10.

The input waveform timing is referenced to the input signal crossing the VREF level applied to the device. The device must still meet the VIH(a) and VIL(ac) specifications under actual use conditions. The minimum slew rate must be maintained in the $V_{ILmax}(ac)$ and $V_{ILmin}(ac)$ signal swing range

8.4.5   *Power and Heat Management*

Power management for very fast integrated circuits is frequently a matter of adequate cooling techniques. Ceramic packages are used in mild situations. For more extreme case, cooling towers can be found on top of packages.

Table 8.10   Ac test conditions for the SSTL2 interface

V_{REF}	Input reference voltage	0.5 *VDDQ
V_{swing}	Input signal maximum peak to peak swing	1.5 V
SLEW	Input signal minimum slew rate	1.0 V/ns

Bibliography

1. Salters, R. and Prince, B., IC's going on a 3V diet, *IEEE Spectrum*, May 1992.
2. Invited talk on basics of transmission lines at JEDEC JC16 meeting, February 1991, Bruce Wenniger.
3. Quinnell, R.A., High-speed bus interfaces, *EDN*, 30 September 1993, 43.
4. Foss, R.C., and Prince, B., Fast Interfaces, *IEEE Spectrum*, October 1992, 54.
5. *Memory Products Databook, SRAMs, ASMs, EEPROMs*, 1993.
6. Motorola, *Fast Static RAM Databook*, 1993.
7. *Interface Standard for Nominal 3V/3.3V Suppy Digital Integrated Circuits*, EIA JEDEC Standard 8-A, June 1994.
8. Klaus, A., Output loading effects on fast static RAMs, AN 1243, *Motorola Fast Static RAM Components and Module Data book*.
9. Bruce Wenniger, private correspondence.
10. Fujitsu, 16M Synchronous DRAM Datasheet.
11. *Standard for Operating Voltages and Interfac Levels for Low Voltage Emitter-Coupled Logic (ECL) Integrated Circuits*, EIA JEDEC Standard 8-2, March 1993.
12. *Gunning Tranceiver Logic (GTL) Low level, High Speed Interface Standard for Digital Integrated Circuits*, EIA JEDEC Standard 8-2, November 1993.
13. *Center-Tap-Terminated (CTT) Low-Level, High Speed Interface Standard for Digital Integrated Circuits*, EIA JEDEC Standard 8-4, November 1993.
14. *2.5V +/− V (Normal Range) and 1.8V to 2.7V (Wide Range) Power Supply Voltage and Interface Standards for Nonterminated digital Integrated Circuits*, EIA JEDEC Standard 8-5, October 1995.
15. *High Speed Transceiver Logic (HSTL)*, EIA JEDEC Standard 8-6, August 1995.
16. *1.8V +/− (Normal Range), and 1.2–1.95 V (Wide Range) Power Supply Voltage and Interface Standard for Non-Terminated Digital Integrated Circuits*, EIA JEDEC Standard 8-7, February, 1997.
17. *Stub Series Terminated Logic for 3.3 volts (SSTL-3)*, EIA JEDEC Standard 8-8, August 1996.
18. *Stub Series Terminated Logic for 2.5 Volts (SSTL-2)*, EIA JEDEC Standard 8-9, September 1998.
19. *Direct Rambus Datasheet*, Rambus Inc. 1998.
20. T. Powell, *et al.*, A 256M SDRAM BIST for disturb test application, *International Test conference*, 1997, pp.200.

9 Fast Packaging Techniques

9.1 Fast Memory Component Packaging

Memory packaging has evolved from thick dual-in-line packages with 100 mil lead centers which put a few megabytes of memory onto foot-square printed circuit boards, to hand-sized SIMM modules that install tens of megabytes of memory into a desktop computer, to multichip modules with a few hundred microns of spacing between the chips that put megabytes of memory into a handheld computer.

Along the way the memories have gotten faster and less tolerant of high package capacitance and inductance that tend to slow them down. Along with the miniature packages have evolved the assembly equipment needed to build and install them in systems and the computer simulation methods required to ascertain their electrical characteristics in the system.

9.2 Packages for Fast DRAMs

9.2.1 Trends to Smaller Sizes in Commodity DRAM Packages

DRAM package sizes used to be determined by the number of required pins on the part. This was particularly true for the older dual-in-line packages which had the pins with 2.54 mm (100 mil) centers. This meant that if the part required 28 pins, the package needed to be $14 \times 2.54 = 3.56$ cm in length. The 28-pin DIP, for example, had 28 pins with 2.54 mm pin centers and was 3.56 cm long and 1.34 cm wide. The length of the package was fixed, so as the DRAM chip was reduced in size over time, it was the width of the package that decreased in size.

Speed was frequently compromised in the attempts to reduce the size of the DRAM packages. For example, the addresses on the DRAM were multiplexed so that the number of pins could be reduced and the package size therefore decreased.

Multiple power and ground pins, required for high speed operation, would have been impossible in a package with 2.54 mm pin centers housing a multi-megabyte density DRAM.

Table 9.1 shows the package size and lead spacing for three 28 pin DRAM packages, the DIP, the SOJ and the TSOPII.

The 28-lead SOJ is only half the length of the comparable 28-pin DIP since the pin centers are 1.27 mm or half the 2.54 mm spacing of the DIP. Packages with smaller lead spacings and thinner leads have decreased the capacitance of the package which in turn permits faster operation.

As assembly technology permitted packages with smaller lead spacings to be used, it has also been possible to add pins to the DRAMs which enhance the speed of the part such as the additional power and ground pins on the SDRAM pinout. For example the body size of the 28-pin TSOPII and the 44-pin TSOPII , shown in Table 9.1, are the same because the pin centers are again halved.

Table 9.1   Dimensions and lead spacing for various DRAM packages

Pins	Package type	Length (mm)	Width (mm)	Pin centers (mm (mils))
28	DIP	35.6	13.4	2.54 (100)
28	SOJ	18.2	10.2	1.27 (50)
28	TSOPII	18.3	10.2	1.27 (50)
44	TSOPII	18.4	10.2	0.9 (35)

9.2.2   Reverse Pinout Packages for Double Sided Modules

The use of modules for assembly of DRAMs permits the DRAMs to be mounted back-to-back. Modules with closely spaced DRAMs or back-to-back mounted DRAMs with short connections through the module can run faster than boards with the DRAMs widely spaced.

To facilitate connected back-to-back mounting, a reverse bend package has been developed with a reverse pinout. An example of the reverse pinout is shown in Figure 9.1 for a 1M×16 DRAM in TSOPII [1].

9.2.3   Vertical DRAM Packages

Vertical packages have also been developed for DRAMs in both the through-hole type and the surface mount package type. The Zigzag-in-line (ZIP) package was developed and used extensively for through-hole mounting. An example of a 4M DRAM in ZIP package is shown in Figure 9.2. The pinout, side view and end view are shown in Figures 9.2(a), (b) and (c) respectively.

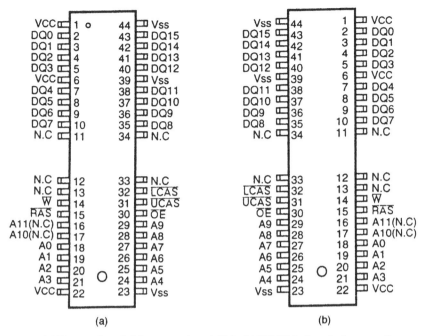

Figure 9.1   (a) Forward and (b) reverse bend 1M×16 DRAM pinouts (source: Samsung [1])

The most familiar vertical surface mount package (VSMP) is that used for the first generation Rambus DRAMS and the SLDRAM. The top view of the pin assignment for the SLDRAM is shown in figure 9.2(d).

There was some early difficulty in stabilizing the vertical surface mount packages, due to the fine lead thickness and lead spacing required in packaging of high density fast DRAMs.

9.2.4   *Speciality DRAM Packages*

VDRAM packaging

Extra pins appear on the VDRAM due to the addition of the serial port. This increases the size of the package over that of the commodity DRAM which both increases the cost of the package and requires additional space on the printed circuit board.

For example, a 1Mb DRAM organized 256K×4 has 20 pins which include nine address pins and four multiplexed input/output pins; while a 1Mb 256×4 multiport video DRAM includes, in addition, four serial input/output and four control pins resulting in a 28-pin package.

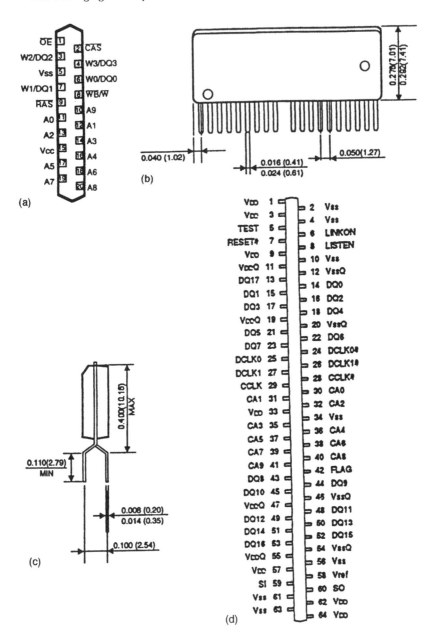

Figure 9.2   4M DRAM in ZIP package (source: Samsung [1]), (d) Top view of the pinout of a 64-pin VSMP (source: JEDEC Standard 21C)

Field memory packaging

Serial DRAMs used as field memories also have a special package since the serial inputs eliminate the address pins and permit a smaller, more cost-effective package.

For example, a 1Mb 256K×4 field memory (TI) eliminates the nine address pins of the 1Mb DRAM and adds four pins to demultiplex the input/output ports resulting in only 16 pins compared to the 20 on the 1Mb DRAM (plus one extra control pin) and compared to 28 on the multiport VDRAM.

Chip scale packaging

DRAMS are now beginning to appear in the Ball Grid Array (BGA) package. This package was originally used for SRAMs and is described in a later section for SRAMs. The BGA package is mounted with balls on the bottom of the package rather than with a lead frame on the sides of the package. For this reason the footprint can be fixed and the external dimensions of the package vary with chip size. BGA packages with this property are called Chip Scale packaging (CSP).

The Direct Rambus DRAM is one of the first of the high speed DRAMs to be offered in a BGA package.

9.3 DRAM SIMM and DIMM Modules

The trend to smaller DRAM packages has been matched by a trend to smaller printed circuit boards for mounting the DRAMs in the system. These small printed circuit boards, or modules, have also been standardized to give the benefits of interchange-ability in the system and cost reduction from mass production.

9.3.1 8/9 Bit SIMM Module

DRAM modules are offered in configurations that permit the entire main memory of a system to be installed on one vertical board. This not only takes up a smaller amount of board space in the system, but permits modules to be treated as standard components.

By mounting the DRAMs relatively close together the modules can also keep the speed characteristics of the module close to those of a single DRAM. Figure 9.3(a) illustates a single-in-line module with eight individual 4M×1 DRAM packages mounted on it in a 4M×8 configuration. It is a convenient way to add 4MB of 5 V memory to a small computer system.

This module can be exchanged for one with identical functionality constructed of two 4M×4 DRAMs as shown in Figure 9.3(b). Externally the two modules are iden-tical. Both modules can be specified with a 60 ns read access time. Their power consumption at this speed, however, differs significantly.

Power consumption for a DRAM module is roughly proportional to the number of DRAM packages on the module. For the first module with the eight 60 ns read access time 4M DRAMs cycling in a Read, Write, or CBR mode, the active current is 840 mA

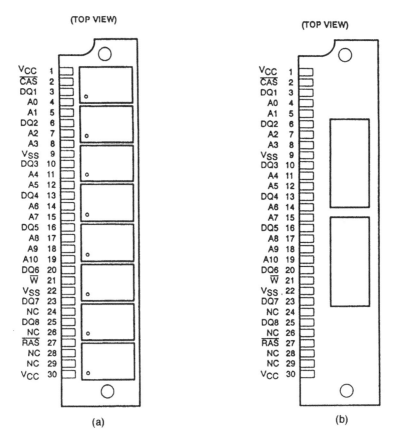

Figure 9.3 4M×8 single-in-line DRAM modules: (a) made with eight 4M×1 DRAMS; (b) made with two 4M×4 DRAMs (source: Texas Instruments [2])

which is 4.2 W at 5 V. The active current of the second module with the two 60 ns 16M DRAMs becomes 220 mA or 1.1 W.

Not only does this power dissipation heat up the part and shorten its life, it also slows it down. Figure 9.4 shows the access time from RAS\ and CAS\ vs. temperature for a typical 4M DRAM [7].

The module with the smaller number of DRAM packages, as well as consuming less power, would also run cooler and, therefore, somewhat faster.

It would also be expected to be a speed upgrade if the level of technology of the 16M DRAM on the upgraded module was higher than that of the 4M DRAMs used on the original module. This is not always the case, however, since both parts could be run in the same process technology.

Note that the connector pinout on this 4M×8 module has two sets of power and ground pins for eight I/O pins or one power or ground pin for every two outputs. This ratio is considered good practice for a quiet memory system running at reasonable speeds.

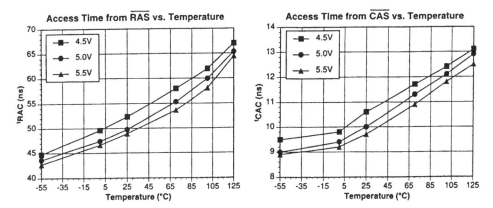

Figure 9.4 Typical 4M DRAM access time from RAS\ and CAS\ v. temperature (source: Micron Technology)

These ×8 or ×9 modules can actually be used in higher speed systems by interleaving them as separate memory banks, thereby effectively doubling the speed of the memory system. The two-bank synchronous DRAM essentially duplicates this performance on a single memory chip, at a higher speed since the buffer logic is brought onto the DRAM chip. Four ×8 modules can also be used on a 32-bit bus to increase the bandwidth.

9.3.2 "×32" SIMM Modules (72-Pin SIMM)

For systems with a 32-bit bus, a single 72-pin module can be used. This module can be constructed as 1M×32, 2M×32, 4M×32, and 8M×32.

The 2M depth module can be attained from the 1M depth module by using DRAMs with a reverse (mirror image) pinout on the opposite side of the board, giving what is called a "double-sided" module. An example of a direct and reverse pinout DRAM package was shown in Figure 9.2. The two parts can be connected through the module board so that the short connections allow the depth to be doubled with no loss in speed for the memory.

A top and side view of several configurations of a 72-pin ×32 module are shown in Figure 9.5.

The 1M×32 is constructed of eight 1M×4 DRAMs mounted on one side of a SIMM module as shown in Figure 9.5(a) and (b). The 2M×32 module is a double-sided SIMM module with 16 1M×4 DRAMs mounted as shown in Figure 9.5(c). Eight of the 1M×4 DRAMs have direct rotation pinouts and are on the front side. Eight have reverse rotation pinouts and are on the back side.

The active current at 60 ns of the 1M×32 made with eight 1M×4 DRAMs is 960 mA or 4.8 W, while for the 2M×32 it is 976 mA, only increasing to 5 W. The reason for the minimal increase in power dissipation is that the majority of the power dissipated in a single DRAM package is in the DRAM sense amplifiers. Since increasing the depth of the memory module does not increase the number of sense amplifiers on at one time, the increase in power dissipation is minimal.

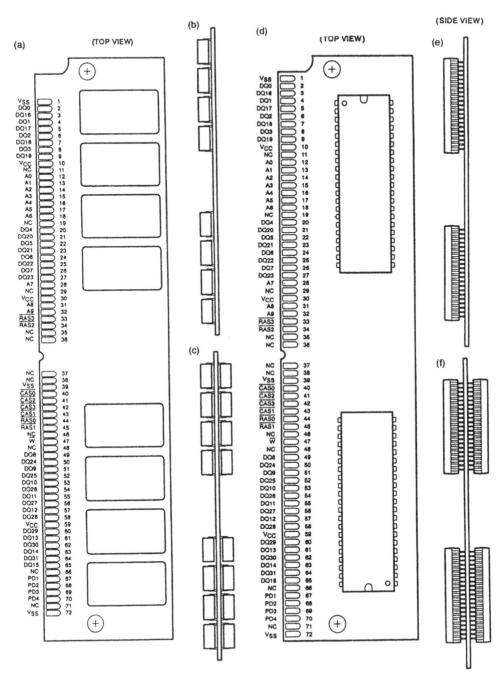

Figure 9.5 Top and side view of several 72-pin (×32) SIMM modules: (a) 1M×32 module made with 4M DRAMs (top view); (b) 1M×32 module made with eight 1M×4 DRAMs (side view); (c) 2M×32 module made with 16 1M×4 DRAMs (side view); (d) 1M×32 module made with 16M DRAMs (top view); (e) 1M×32 module made with two 1M×16 DRAMs (side view); (f) 2M×32 module made with four 1M×16 DRAMs (side view) (source: Texas Instruments)

As in the case of the ×8 module, it is possible to reduce the package count and hence the power consumption for a 1M×32 module by replacing the eight 1M×4 DRAMs with two 1M×16 DRAMs mounted on one side of the SIMM as shown in Figure 9.5(d) and (e). The active current at 60 ns read access using 16M DRAMs is 390 mA or 1.95 W, less than half of the 4.8 W of the 4M DRAMs.

As before, a 2M×32 module can be made using the 16M DRAM by making the module double sided. Two reverse pinout 1M×16 DRAMs are mounted on the back side of the module as shown in Figure 9.5(f).

Since the module is only being increased in depth, it would be expected that the active power dissipation would not increase significantly. The active current at the same speed is 394 mA or 1.97 W.

Since the 2M×32 module made with 16M DRAMs has a lower power dissipation than that of the 4M DRAMs, we would also expect that in the same technology the 16M DRAM module would run faster.

The 72-pin SIMM has only three sets of power and ground pins for 32 outputs. This is less than one power or ground for five outputs which is less than expected in normal good design practice for a fast module to prevent ground bounce noise. This module was created for commodity PC systems and was never expected to run faster than 40 ns fast page mode or about 25 MHz.

The 72-pin SIMM created for the 4M DRAM has wasted board area when used with the 16M DRAMs as could be seen in Figure 9.5. The area was actually required for the 72-pin SIMM connector. Since the SIMM connector has the same pins on both sides of the SIMM, it was a small step to the next DRAM module concept, the DIMM, which folds the SIMM connector in half.

9.3.3 Small Outline 72-Pin DIMMs

A shorter module has been defined using distinct dual-in-line connectors on the two sides of the board which essentially folds the 72-pin SIMM module in half. This "dual-in-line" SIMM is called a DIMM. Using the concept of a DIMM module, it is possible to have a small 72-pin connector by having 36 of the pins on one side of the module and the other 36 on the other side. Such a module can still be made single-sided or double-sided to increase the depth of the memory.

Figures 9.6(a) and 9.6(b) show diagrams of a 1M×32 72-pin Small Outline DIMM (SO DIMM) module made single-sided with two 1M×16 DRAMs on one side of the package. The top view is shown in Figure 9.6(a) and the side view in Figure 9.6(b).

The power dissipation of a SO DIMM also could have been a problem if operated at 5 V since the same amount of heat is being dissipated in a smaller board area. For this reason, the SO DIMMs were primarily offered with 16M DRAMs with 3.3 V power supplies which reduced the power dissipation level. For example, a 2M×32 SO DIMM at 3.3 V has an active power dissipation of 0.5 W whereas a 2M×32 SIMM at 5 V has an active power dissipation of 1.95 W. There is reduction by a factor of four in going to the lower power supply voltage.

There is some flexibility in building these modules. For example a 2M×32 SO DIMM can be built as a double-sided module with four 1M×16 DRAMs back-to-back as shown in Figure 9.6(c), or it can be built as a single sided module with four 2M×8 DRAMs on one side.

The difference between the two types of 2M×32 SO DIMMs is in the power dissipation, refresh, and speed. These are compared in the following chart:

	4× (1M×16)	4×(2M×8)
Refresh	4096	2048
Addressing	11/10	12/8
ICC1 (60)	190	480

Proceeding to larger modules, it is possible for a 4M×32 DIMM module to be made from a 2M×32 module made with four 2M×8 DRAMs on one side by adding four more 2M×8 DRAMs back-to-back using reverse pinout DRAMs to form a double-sided module.

A 4M×32 module can also be made single-sided with DRAMs mounted on both sides by using eight 4M×4 DRAMs four to a side, showing that it is also possible to make a DIMM single-sided module with units mounted on both sides of the package. In this case reverse pinned units are not used and the module is the same depth as the DRAMs.

The 72-pin SIMM or SO DIMM supports systems with 32-bit buses. As systems with 64-bit buses appeared, a new module was standardized, the 168-pin DIMM.

9.3.4   168-Pin 64/72-Bit (8-Byte) DRAM DIMM Module

A faster 168-pin DIMM module has been created to support the 64-bit bus of the 586 class processors. This module is also predominantly used with 3.3 V 16M density DRAMs in densities ranging from 8MB (1M×64/72) to 64MB (8M×64/72) and with 64M DRAM upgrades to 128MB (16M×72).

Figure 9.7 compares three configurations of this eight-byte module.

A 4M×72 DIMM module made with 18 4M×4 DRAMs mounted on both sides of a single-sided module is shown in Figure 9.7(a). If the same DIMM configuration is populated with an 8M×4 component, made of two 4M (4M×4) DRAM components stacked one on top of the other, then we have an 8M×72 DIMM module made with 36 4M×4 DRAMs on a single-sided DIMM module (Figure 9.7(b)). Finally an 8M×72 may be made with nine 8M×8 DRAMs on a two sides of a single-sided DIMM module (Figure 9.7(c)) [3].

If we compare the power consumption of the two 8M×72 modules, the power consumption is about half for the 64M DRAM module even though only one-fourth the number of components is used. It is not a fourth since the 4M DRAM module is essentially made from 18 8M double-sided components mounted on a single-sided module.

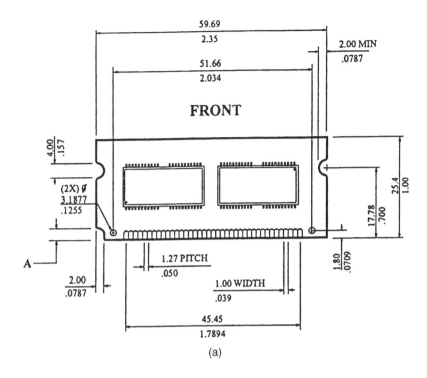

(a)

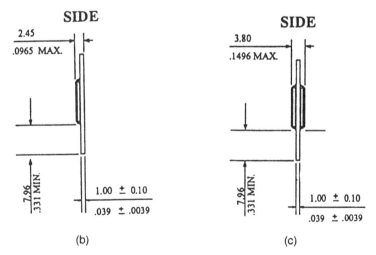

(b)   (c)

Figure 9.6 1MX×32 72-pin small outline DIMM: (a) top view of 1M/2M×32 module (two
1M×16 DRAMs); (b) side view of 1M×32 single-sided (two 1M×16 DRAMs);
(c) 2M×32 double-sided (four 1M×16 DRAMs) (source: IBM [3])

The pinout for the 8M×72 DIMM module is shown in Figure 9.8(a). To reduce
noise there are 32 power or ground pins. That is, one power or ground pin is
provided for every four I/O pins, which is reasonable for a module intended to
run at a fast page mode cycle time of 35 ns (28 MHz).

System performance is improved by having the inputs buffered except for the RAS\ and data. Leaving the data and RAS\ signals unbuffered preserves the 60 ns DRAM access specifications on the module which permits a 35 ns fast page cycle time.

These 168-pin DIMM modules are also made using DRAMs with the newer EDO page mode (Hyperpage). The same 60 ns access time module which would have supported a 35 ns fast page cycle time would support a 25 ns (40 MHz) EDO page mode (Hyperpage) cycle [4].

Systems using these DIMMs operate at 40–50 MHz not because the processor can not operate at higher clock speeds, but because the transition to higher speeds at the board level is difficult.

The memory modules suffer from reflections and cross talk. Every connection and change of direction of a trace (wire) causes reflections. Adjacent wires with signals traveling in the same direction without sufficient shielding can cause electromagnetic interference, called "cross-talk", with each other. Cross-talk can also slow down the effective rate of the system.

As faster DRAMs, such as the synchronous DRAM, were developed, a suitable module layout was needed for mounting the faster DRAM. The original synchronous DRAMs operate up to 100 MHz and faster ones follow. A module for these faster DRAMs would need good layout backed up by relevant transmission line simulation models to compensate for clock skew across the module. Connectors adequate for a 100 MHz speed would also be required, as would fast logic chips.

The first SDRAM module was an adaptation of the JEDEC Standard 168-pin DIMM module. This module had the advantage of minimal tooling changes being required to adapt from the DRAM to the SDRAM. Its disadvantage was the insufficient number of pins available to adequately ground the module for very high speed SDRAMs. Since the first synchronous DRAMs were used at speeds less than 100 MHz, this DIMM module was adequate.

A pinout of a 168-pin SDRAM DIMM module is shown in Figure 9.8(b). This pinout is for the PC100 DIMM which was used widely with 100 MHz SDRAMS for personal computer main memory. Figure 9.8(c) shows the top view of the pin assignment of the connector into which the vertically mounted DIMM fits.

Notice that both the DRAM and SDRAM 168-pin DIMM modules have power and ground pins for every four DQ pins.

The external characteristic identification for the early DIMM packages was handled by adding an additional pin for each characteristic it was necessary to determine for the DIMM. This included such items as: manufacturer, organization of DIMM, speed of DIMM, etc. As the list of characteristics the user wanted to electrically determine grew, it became unfeasible to add a pin for each.

The solution for electrically providing multiple identifying characteristics for a DIMM package was solved by changing from the parallel presence detect method, which added a pin for each characteristic, to a serial presence detect method. The Serial Presence Detect (SPD) maintains the same number of pins regardless of the amount of information stored.

An example of 64 identifying characteristics that are stored in the SPD for a PC100 168-pin SDRAM DIMM is shown in the chart in Figure 9.8(d). A specialized EPROM is used to store this information.

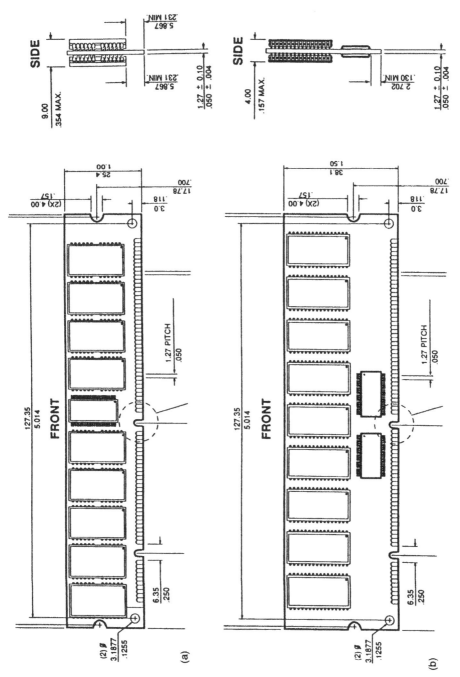

Figure 9.7 168-Pin single-sided DIMM module (top and side view): (a) 4M×72 using 18 4M×4 DRAMs; (b) 8M×72 using 36 stacked 4M×4 DRAMs; (c) 8M×72 using nine 8M×8 DRAMs (source: IBM [3])

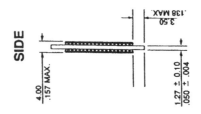

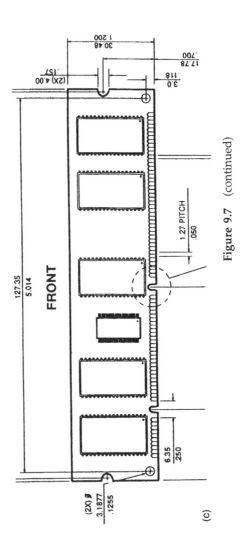

Figure 9.7 (continued)

Pin#	Front Side	Pin#	Back Side	Pin#	Front Side	Pin#	Back Side
1	Vss	85	Vss	43	Vss	127	Vss
2	DQ0	86	DQ36	44	$\overline{OE2}$	128	NC
3	DQ1	87	DQ37	45	$\overline{RAS2}$	129	$\overline{RAS3}$
4	DQ2	88	DQ38	46	$\overline{CAS4}$	130	NC
5	DQ3	89	DQ39	47	NC	131	NC
6	Vcc	90	Vcc	48	$\overline{WE2}$	132	$\overline{PDE}$
7	DQ4	91	DQ40	49	Vcc	133	Vcc
8	DQ5	92	DQ41	50	NC	134	NC
9	DQ6	93	DQ42	51	NC	135	NC
10	DQ7	94	DQ43	52	DQ18	136	DQ54
11	DQ8	95	DQ44	53	DQ19	137	DQ55
12	Vss	96	Vss	54	Vss	138	Vss
13	DQ9	97	DQ45	55	DQ20	139	DQ56
14	DQ10	98	DQ46	56	DQ21	140	DQ57
15	DQ11	99	DQ47	57	DQ22	141	DQ58
16	DQ12	100	DQ48	58	DQ23	142	DQ59
17	DQ13	101	DQ49	59	Vcc	143	Vcc
18	Vcc	102	Vcc	60	DQ24	144	DQ60
19	DQ14	103	DQ50	61	NC	145	NC
20	DQ15	104	DQ51	62	NC	146	NC
21	DQ16	105	DQ52	63	NC	147	NC
22	DQ17	106	DQ53	64	NC	148	NC
23	Vss	107	Vss	65	DQ25	149	DQ61
24	NC	108	NC	66	DQ26	150	DQ62
25	NC	109	NC	67	DQ27	151	DQ63
26	Vcc	110	Vcc	68	Vss	152	Vss
27	$\overline{WE0}$	111	NC	69	DQ28	153	DQ64
28	$\overline{CAS0}$	112	NC	70	DQ29	154	DQ65
29	NC	113	NC	71	DQ30	155	DQ66
30	$\overline{RAS0}$	114	$\overline{RAS1}$	72	DQ31	156	DQ67
31	$\overline{OE0}$	115	NC	73	Vcc	157	Vcc
32	Vss	116	Vss	74	DQ32	158	DQ68
33	A0	117	A1	75	DQ33	159	DQ69
34	A2	118	A3	76	DQ34	160	DQ70
35	A4	119	A5	77	DQ35	161	DQ71
36	A6	120	A7	78	Vss	162	Vss
37	A8	121	A9	79	PD1	163	PD2
38	A10	122	A11	80	PD3	164	PD4
39	NC	123	NC	81	PD5	165	PD6
40	Vcc	124	Vcc	82	PD7	166	PD8
41	NC	125	NC	83	ID0	167	ID1
42	NC	126	B0	84	Vcc	168	Vcc

(a)   **Note:** All pin assignments are consistent for all 8 Byte versions.

Figure 9.8   Illustration of pinouts for 168-pin DIMM modules: (a) pinout for a 8M×72 DIMM module source IBM [3]); (b) pinout for a 16M×64 SDRAM DIMM module source: Toshiba [11]; (c) top view of connector pinout for 168-pin DIMM module source: Toshiba [11]; (d) serial presence detect for the PC100 168-pin DIMM module (source: Toshiba [11])

1	VSS	85	VSS	29	DQMB1	113	DQMB5	57	DQ18	141	DQ50
2	DQ0	86	DQ32	30	/CS0	114	/CS1	58	DQ19	142	DQ51
3	DQ1	87	DQ33	31	NC	115	/RAS	59	VDD	143	VDD
4	DQ2	88	DQ34	32	VSS	116	VSS	60	DQ20	144	DQ52
5	DQ3	89	DQ35	33	A0	117	A1	61	NC	145	NC
6	VDD	90	VDD	34	A2	118	A3	62	NC	146	NC
7	DQ4	91	DQ36	35	A4	119	A5	63	CKE1	147	NC
8	DQ5	92	DQ37	36	A6	120	A7	64	VSS	148	VSS
9	DQ6	93	DQ38	37	A8	121	A9	65	DQ21	149	DQ53
10	DQ7	94	DQ39	38	A10	122	BA0	66	DQ22	150	DQ54
11	DQ8	95	DQ40	39	BA1	123	A11	67	DQ23	151	DQ55
12	VSS	96	VSS	40	VDD	124	VDD	68	VSS	152	VSS
13	DQ9	97	DQ41	41	VDD	125	CLK1	69	DQ24	153	DQ56
14	DQ10	98	DQ42	42	CLK0	126	NC	70	DQ25	154	DQ57
15	DQ11	99	DQ43	43	VSS	127	VSS	71	DQ26	155	DQ58
16	DQ12	100	DQ44	44	NC	128	CKE0	72	DQ27	156	DQ59
17	DQ13	101	DQ45	45	/CS2	129	/CS3	73	VDD	157	VDD
18	VDD	102	VDD	46	DQMB2	130	DQMB6	74	DQ28	158	DQ60
19	DQ14	103	DQ46	47	DQMB3	131	DQMB7	75	DQ29	159	DQ61
20	DQ15	104	DQ47	48	NC	132	NC	76	DQ30	160	DQ62
21	NC	105	NC	49	VDD	133	VDD	77	DQ31	161	DQ63
22	NC	106	NC	50	NC	134	NC	78	VSS	162	VSS
23	VSS	107	VSS	51	NC	135	NC	79	CLK2	163	CLK3
24	NC	108	NC	52	NC	136	NC	80	NC	164	NC
25	NC	109	NC	53	NC	137	NC	81	NC	165	SA0
26	VDD	110	VDD	54	VSS	138	VSS	82	SDA	166	SA1
27	/WE	111	/CAS	55	DQ16	139	DQ48	83	SCL	167	SA2
28	DQMB0	112	DQMB4	56	DQ17	140	DQ49	84	VDD	168	VDD

(b)

(c)

Figure 9.8   (*continued*)

9.3.5   *72-Bit (8-Byte) 200-Pin Synchronous DRAM DIMM Module*

An eight byte DIMM module has also been developed to optimize for the higher speed synchronous DRAM. The pinout of this 200-pin module is shown in Figure 9.9. The component used in this module is the 2M×8 SDRAM.

This module is intended to run at burst speeds of 80 MHz to 166 MHz and above. Because of the higher speed, one power or ground pin is provided for every two I/O's to reduce ground bounce. The module is configured using 16M (1M word × 8 bit × 2 bank) SDRAMs. Two 16-bit registers are provided on the module to buffer all inputs except the I/O pins. A 12-bit Phase Lock Loop (PLL) is provided on the clock (CLK0) input to synchronize the signals to or from the various chips and reduce chip skew.

Byte Number	Function Described	-80		-10	
		Entry Value	Entry	Entry Value	Entry
0	Defines # Bytes Written into Serial Memory at Module mfgr	128 Bytes	80h	128 Bytes	80h
1	Total # Bytes in SPD Memory Device	256 Bytes	08h	256 Bytes	08h
2	Fundamental Memory Type (FPM, EDO, SDRAM...) from Appendix A	SDRAM	04h	SDRAM	04h
3	# Row Addresses on this Assembly	RA0-RA11	0Ch	RA0-RA11	0Ch
4	# Column Addresses on this Assembly	CA0-CA8	09h	CA0-CA8	09h
5	# Module Banks on this Assembly	2 Bank	02h	2 Bank	02h
6	Data Width of this Assembly...	x64	40h	x64	40h
7	...Data Width Continuation	x64	00h	x64	00h
8	Voltage Interface Standard for this Assembly	LVTTL	01h	LVTTL	01h
9	SDRAM Cycle Time at Max. Supported CAS Latency (CL), CL = X	CL = 3, 8.0 ns	80h	CL = 3, 10 ns	A0h
10	SDRAM Access from Clock @ CL = X	CL = 3, 6.0 ns	60h	CL = 3, 7.0 ns	70h
11	DIMM Configuration Type (Non-parity, Parity, ECC)	Non/Parity	00h	Non/Parity	00h
12	Refresh Rate/Type	15.625 µs/Self-Refresh	80h	15.625 µs/Self-Refresh	80h
13	SDRAM Width, Primary DRAM	x8	08h	x8	08h
14	Error Checking SDRAM Data Width	x8	08h	x8	00 08h
15	Minimum Clock Delay, Back-to-Back Random Column Addresses	1 CLK	01h	1 CLK	01h
16	Burst Lengths Supported	1,2,4,8 Full page	8Fh	1,2,4,8 Full page	8Fh
17	# Banks on Each SDRAM Device	4 Banks	04h	4 Banks	04h
18	CAS # Latencies Supported	2,3	06h	2,3	06h
19	CS # Latency		01h		01h
20	WE # Latency		01h		01h
21	SDRAM Module Attributes		00h		00h
22	SDRAM Device Attributes: General		0Eh		0Eh
23	Minimum Clock Cycle Time at CL- X-1	CL = 2, 10 ns	A0h	CL = 2, 12 ns	C0h
24	Maximum Data Access Time from Clock @ CL X-1	CL = 2, 6.0 ns	60h	CL = 2, 8.0 ns	80h
25	Minimum Clock Cycle Time at CL X-2		00h		00h
26	Maximum Data Access Time from Clock @ CL X-2		00h		00h
27	Minimum Row Precharge Time	20 ns	14h	24 ns	18h
28	Minimum Row-Active-to-Row-Active Delay	20 ns	14h	20 ns	14h
29	Minimum RAS-to-CAS Delay	20 ns	14h	24 ns	18h
30	Minimum RAS Pulse Width	48 ns	30h	60 ns	3Ch
31	Module/Bank Density	64 MB	10h	64 MB	10h
32	Command & address signal input Setup Time	2 ns	20 h	2.5 ns	25 h
33	Command & address signal input Hold Time	1 ns	10 h	1 ns	10 h
34	Data Signal input Setup Time	2 ns	20 h	2.5 ns	25 h
35	Data Signal input Hold Time	1 ns	10 h	1 ns	10 h
36-61	Superset Information (May Be Used in Future)		FFh		FFh
62	SPD Revision	Rev. 1.2A	12h	Rev. 1.2A	12h
63	Checksum for Bytes 0-62	1ED2h	D2h	1F60h	60h

Option

64	Manufacturers JEDEC ID Code per JEP-106E				
65-71					
72	Manufacturing Location				
73-90	Manufacturer's Part Number				
91-92	Revision Code				
93-94	Date of Manufacture				
95-98	Assembly Serial Number				
99-125	Manufacturer-Specific Data				
126	Reserved	Intel Specification	66h	Intel Specification	66h
127	Reserved	Intel Specification	F7h	Intel Specification	F7h
128-255					

(d)

Figure 9.8 (*continued*)

The layout of the SDRAM DIMM module is shown in the block diagram in Figure 9.10 [5].

This DIMM is laid out to minimize the skew across the module by equalizing the wiring length and having the input/output and phase lock loop in the center of the package.

Pin Arrangement

Pin No.	Pin Name	Pin No.	Pin. Name	Pin No.	Pin Name	Pin No.	Pin Name
1	VDD	51	VSS	101	NC	151	CLK0
2	NC	52	/RE	102	NC	152	VDD
3	NC	53	VSS	103	VSS	153	/CS1
4	NC	54	NC	104	NC	154	/CS0
5	NC	55	NC	105	NC	155	VSS
6	ID0	56	VDD	106	CLR	156	NC
7	ID1	57	A0	107	ID2	157	A10
8	VSS	58	A1	108	DQ71	158	VDD
9	DQ67	59	VSS	109	DQ70	159	A2
10	DQ66	60	DQ35	110	VSS	160	A3
11	VDD	61	DQ34	111	DQ69	161	VSS
12	DQ65	62	VDD	112	DQ68	162	DQ31
13	DQ64	63	DQ33	113	VDD	163	DQ30
14	VSS	64	DQ32	114	NC	164	VDD
15	DQ63	65	VSS	115	VSS	165	DQ29
16	DQ62	66	DQ27	116	NC	166	DQ28
		67	DQ26			167	VSS
Interface key		68	VDD	Interface key		168	DQ23
		69	DQ25			169	DQ22
		70	DQ24			170	VDD
17	NC	71	VSS	117	DQ59	171	DQ21
18	DQ61	72	DQ19	118	DQ58	172	DQ20
19	DQ60	73	DQ18	119	VSS	173	VSS
20	VDD	74	VDD	120	DQ57	174	NC
21	NC	75	DQ17	121	DQ56	175	NC
22	NC	76	DQ16	122	VDD	176	VDD
23	VSS	77	VSS	123	DQ55	177	NC
24	NC	78	NC	124	DQ54	178	VSS
25	NC			125	VSS		
26	VDD	Voltage key		126	DQ53	Voltage key	
27	DQ51			127	DQ52		
28	DQ50			128	VDD		
29	VSS	79	NC	129	DQ47	179	VSS
30	DQ49	80	VDD	130	DQ46	180	NC
31	DQ48	81	DQ15	131	VSS	181	NC
32	VDD	82	DQ14	132	DQ45	182	VDD
33	DQ43	83	VSS	133	DQ44	183	DQ11
34	DQ42	84	DQ13	134	VDD	184	DQ10
35	VSS	85	DQ12	135	DQ39	185	VSS
36	DQ41	86	VDD	136	DQ38	186	DQ9
37	DQ40	87	DQ7	137	VSS	187	DQ8
38	VDD	88	DQ6	138	DQ37	188	VDD
39	A4	89	VSS	139	DQ36	189	DQ3
40	A5	90	DQ5	140	VDD	190	DQ2
41	VSS	91	DQ4	141	A6	191	VSS
42	A8	92	VDD	142	A7	192	DQ1
43	A9	93	/PDE	143	VSS	193	DQ0
44	VDD	94	PD0	144	A11	194	PD4
45	CKE1	95	PD1	145	NC	195	PD5
46	CKE0	96	PD2	146	VDD	196	PD6
47	VSS	97	PD3	147	DOM	197	PD7
48	/CE	98	NC	148	/WE	198	VDQ
49	NC	99	NC	149	VSS	199	NC
50	VDD	100	VSS	150	NC	200	NC

Figure 9.9   Pinout of 200-pin SDRAM DIMM module (source: Hitachi)

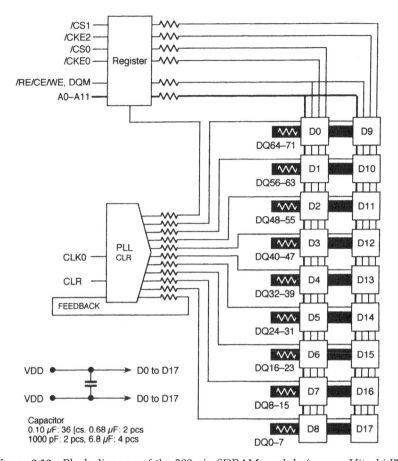

Figure 9.10   Block diagram of the 200-pin SDRAM module (source: Hitachi [5])

The active burst operating current for a two cycle CAS\ latency at 80 MHz operation is specified at 1.0 mA (3.3 W at 3.3 V) and for a four cycle latency at 2.26 A (7.5 W). It is assumed that the outputs are unloaded, and any load would reduce the speed. Input capacitance is specified at 10 pF and output capacitance at 20 pF [5].

It appears that for even faster DRAM modules in the 150–200 MHz range much more effort will be needed. A simple doubling of the speed can of course be effected by sampling on both the rising and falling edges of the clock as is done with the DRAM modules from Rambus, Inc. In this case a 250 MHz clock speed becomes a 500 MHz sample rate. Both a special bus and high speed interface are required to reach the 250 MHz clock speed.

The 200-pin DIMM module is used in some cases for the Double Data Rate (DDR) SDRAM which is expected to clock data at up to 300 MHz. The ground bounce characteristics are improved by the use of the SSTL2 2.5V interface with 64M or 256M DRAM chips mounted on the module.

9.4   Fast SRAM Packages

While the DRAM tends to have a few standard packages which are used in very high volume, the fast SRAM is more applications oriented and has a much wider selection of package types. These range from the ceramic packages and packages with cooling towers used with very high speed bipolar memories to the standardized packages used for the high density fast cache SRAMs.

An example of the ceramic flatpaks used for 101K and 100K high speed ECL applications is shown in Figure 9.11. The two configurations shown are a pinout with center power and ground in Figure 9.11(a) and one with end power and ground in Figure 9.11(b) [4].

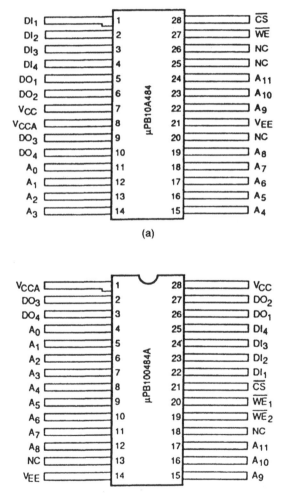

(a)

(b)

Figure 9.11   Ceramic flatpacks used for high speed ECL interfaces: (a) center power and ground; (b) end power and ground

An overview of the variety of packages available for the SRAMs and the trend toward fast packages can be seen by considering the synchronous SRAMs used for cache applications in PCs and workstations. This particular application represents a large part of the fast SRAMs manufactured in the world.

9.4.1 Packages for Fast Synchronous SRAMs

The fast synchronous SRAMs being supplied for PCs are available in densities from 256K up to 1M to 4M. Packages expected to be used in high volume for these SRAMs are the Plastic Leaded Chip Carrier (PLCC), the Thin Quad Flat Pak (TQFP), and the Ball Grid Array (BGA). An estimate of the consumption of these packages from 1995 to 2000 is shown in Figure 9.12 [8].

Current packages for 512K and 1M burst SSRAMs are as shown in Table 9.2. The number of suppliers for each package indicates the variety available.

The major packages for the 1M burst SRAM are the 52-pin PLCC, the 100-pin TQFP, the 119-pin Ball Grid Array, and bare die.

The PLCC was used for the earlier generations of fast cache SRAMs and the first 5 V 64K×18 burst SSRAMs. The TQFP is rapidly becoming the predominant package for the burst SSRAM parts.

The Ball Grid Array is a more compact package. Several suppliers ship the 1M burst SSRAM parts in this package for high performance workstation applications. The bare die option is also offered by several suppliers although all the problems have not been completely overcome. It is used predominantly in high speed modules.

The Plastic Leaded Chip Carrier (PLCC) packages have been on the market longer than the others. The 32K×9, 32K×18 and 64K×18 have all been offered in the PLCC package. Figure 9.13 shows the pinouts of these three parts in the PLCC package.

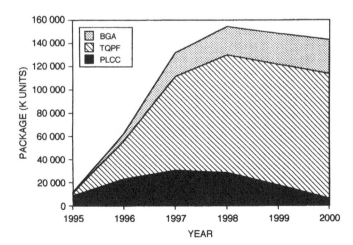

Figure 9.12   1M burst SSRAMs by package (thousand units) (source: Memory Strategies International)

Table 9.2 Packages for burst SSRAMs

	Number of suppliers known to have each package*			
	52PLCC	100TQFP	119BGA	KGD
512K:				
32K×18	3		1	
1M:				
64K×18 5 V	3			
64K×18 3.3 V	6	4	2	3
32K×32 3.3 V		8		2
32K×36 3.3 V	2	3	4	3

*In 1995.

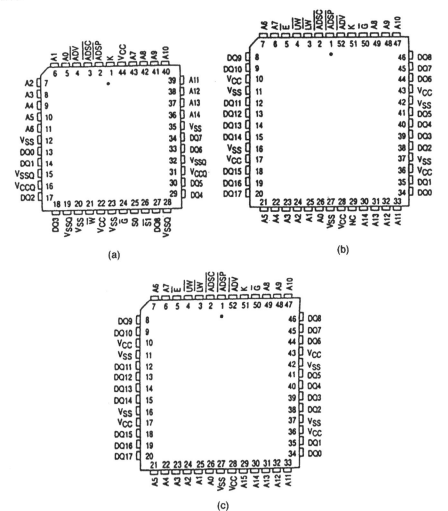

(a)

(b)

(c)

Figure 9.13 Pinouts of the ×9 and ×18 fast cache SRAMs in PLCC package: (a) 32K×9; (b) 32K×18; (c) 64K×18 (source: Motorola [6])

The original 5 V 64K×18 synchronous burst SRAM was supplied in the 52-pin PLCC package because it was the only standard package available at the time this part was introduced and because the larger first-generation chip did not fit in the smaller 100-pin TSOPII package. Another early package for the 32K×32 was a PQFP package. The PQFP is a thicker package with the same pinout as the 100-lead TQFP.

The major package for the 3.3 V 64K×18 burst SSRAM as well as the 32K×32 3.3 V burst SSRAM parts is the 100-pin TQFP which is used in newer PC systems requiring higher speed.

Ball Grid Array

For very fast workstation types of systems, the Ball Grid Array is frequently used. Ball Grid Arrays (BGA), in both ceramic and plastic, can meet both performance and density needs for high pin count chips such as processors and fast SRAMs. The lead solder bumps on the bottom of the BGA are not subject to bent or skewed leads and improve ease of handling. Costs are comparable to that of the QFP.

The 119-pin BGA package is expected to take a share in very high performance systems in the 0.8 mm technology first generation of the 1M burst SSRAMS. It is expected to be more widely supported in the second generation 0.5 micron burst SSRAMs that can use the extra speed opportunity provided by this package.

Figure 9.14(a), (b) and (c) show package top and side diagrams of the 52-pin PLCC, the 100-pin TQFP and the 119-pin BGA respectively [6].

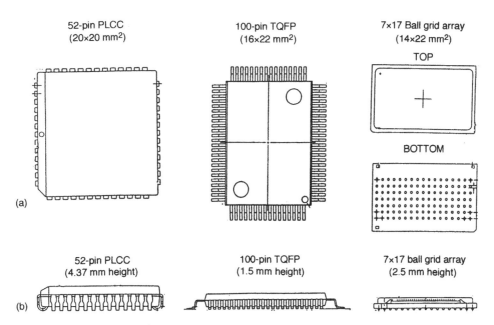

Figure 9.14 Diagrams of various 1M burst SSRAM packages: (a) top views; (b) side views
Source: Motorola [6]

The 1M burst SSRAMs are also offered as bare die in a form referred to as "Known Good Die" by several suppliers.

Some of the module manufacturers, such as Micro-Module Systems, are offering modules which include the Pentium and burst SSRAM in Multi-chip Modules (MCM) using die shipped under KGD programs [9].

A photograph is shown in Figure 9.15 of a fast 512KB SRAM data cache on a multi-chip module composed of fast 1M SRAMs.

9.4.2 Speed Considerations in SRAM Package Selection

The more compact packages for cache SRAMs tend to be faster in the system since they can sit closer to the processor. This is why the integrated P6 processor with the cache SRAM chip in a multichip package can operate at full processor speed.

Multichip modules are the next in order in package speed. These are packaging in which the separate chips have bumped bonding pads and are flipped and mounted onto a substrate with minimal separation between the chips.

The BGA with its low inductance and compact size ranks next if mounted with a processor also in a BGA with minimal spacing. This is followed by the TQFP, the PQFP and the PLCC.

1M cache SRAM packages ranked approximately by speed capability in the system are therefore:

1. Multichip package integrated with processor
2. Multichip modules
3. Ball Grid Array assembly
4. TQFP
5. PQFP
6. PLCC

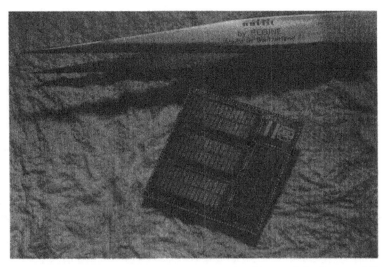

Figure 9.15 Fast SRAM cache subsystem on a multi-chip module Source: Micro-Module Systems [9]

9.4.3   Trends in Systems Using Miniature Packaging

Memory package size has been decreasing because of the need to get faster parts closer together in high performance systems. There has also been a trend for systems to get smaller as higher density chips are available to decrease cost and improve portability. All of these trends work together to promote the smaller package sizes.

Table 9.3 shows the dimensions of the four packages used for the 1M burst SSRAMs from the older 52-pin PLCC, to the 100-pin TSOP and the 119-pin BGA.

The footprint and area occupied for the package differ for the BGA since the bonding pads are underneath the package whereas for the other two packages the bonding pads extend out from the package. As a result of bonding pads beneath the package, the BGA package can extend out beyond the bonding pads and hence occupy more area than the footprint.

Another way to look at the footprint and board area is to note that the board area occupied by the leaded packages is larger than the package dimension since it also includes the space taken up by the leads on the PC board. Given this consideration BGAs save even more board area over leaded packages for sizes greater than 100 leads.

A comparison chart for the BGA and the QFP is shown in Table 9.4.

The speed of the systems has increased and the shrinking package size also permits smaller systems and hence shorter wiring between components, which improves the transmission line characteristics of the system.

Multipackage modules including the processor and the SRAM were the first attempt to bring these components closer together. Multichip packages and modules are the next steps.

Table 9.3   Dimensions of 1M burst SSRAM packages

		52-pin PLCC	100-pin TSOP	199-pin BGA
Footprint	(mm^2)	20×20	22×16	20.3×7.6
	(mm^2)	400	352	155
PC area	(mm^2)	400	352	308

Table 9.4   PC board footprint area for comparable BGA and QFP packages

Package	Pitch (mm)	Total board area (mm^2)
160 QFP	0.65	980
240 QFP	0.5	1197
169 PBGA	1.5	529
225 PBGA	1.5	729
313 PBGA	1.27	1225
256 CBGA	1.27	441

Source: Motorola.

Figure 9.16 plots the trends in package size for the 1M burst SSRAM from the 52-lead PLCC to the bare die.

The BGA offers most of the advantages of bare die with fewer disadvantages. It has a small footprint and short leads to improve speed, yet it can be tested and handled as any other packaged part.

The BGA still suffers from the power dissipation problems faced by all of these fast chips. However, until the bare die solution becomes more of a production process, BGA or some other form of ultra-miniature packaging is likely to remain the volume package of choice for very high speed systems.

9.5 SRAM Modules

9.5.1 Multi-Package SRAM Modules

Since SRAMs have been used for a long time in high speed memory systems and cache subsystems, fast SRAM packages mounted on modules were developed early. These modules permitted the SRAMs to be closer together and therefore run faster.

There has not been the standardization in packaged SRAM modules that has appeared in similar DRAM modules since the volume is lower and the SRAM modules tend to be more applications specific.

An example is a burst SRAM DIMM module for use in Power PC processor systems. The 136-pin DIMM module is shown in Figure 9.17a. Figure 9.17b shows the pinout and 9.17c the pin names. The four SRAMs are packaged in PLCC.

Figure 9.18 shows the 64K×72 module block diagram using four 64K×18 SRAMs.

9.5.2 SRAM Multichip Packages and Multichip Modules

Another applications specific module for SRAMs is the multichip package. This can take two forms: multiple SRAM packages can be mounted on a larger single-chip package, or multiple SRAM bare die can be mounted in a single package.

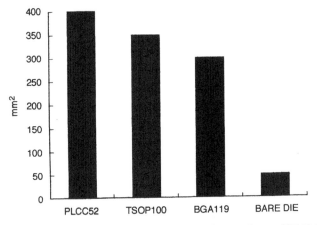

Figure 9.16 Trends in package size for the 1M burst SSRAM

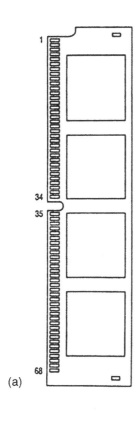

(a)

PD0	1	69	V_{SS}

(b)

PD0	1	69	V$_{SS}$
PD1	2	70	PD2
DQ0	3	71	V$_{CC}$
DQ1	4	72	DQ2
V$_{CC}$	5	73	DQ3
DQ4	6	74	DQ5
DQ6	7	75	DQ7
DQP0	8	76	V$_{SS}$
DQ8	9	77	DQ9
DQ10	10	78	DQ11
V$_{SS}$	11	79	DQ12
K0	12	80	V$_{SS}$
V$_{SS}$	13	81	DQ13
DQ14	14	82	DQ15
V$_{CC}$	15	83	DQP1
DQ16	16	84	V$_{SS}$
DQ17	17	85	DQ18
DQ19	18	86	DQ20
DQ21	19	87	DQ22
V$_{CC}$	20	88	DQ23
DQP2	21	89	V$_{SS}$
DQ24	22	90	DQ25
DQ26	23	91	DQ27
DQ28	24	92	DQ29
V$_{SS}$	25	93	DQ30
DQ31	26	94	V$_{SS}$
DQP3	27	95	$\overline{E0}$
V$_{SS}$	28	96	$\overline{WT}$
$\overline{W0}$	29	97	$\overline{W3}$
$\overline{W2}$	30	98	$\overline{G0}$
$\overline{TSP}$	31	99	$\overline{TSC}$
$\overline{BAA}$	32	100	V$_{SS}$
V$_{CC}$	33	101	$\overline{G1}$
$\overline{W4}$	34	102	$\overline{W5}$
$\overline{W6}$	35	103	$\overline{W7}$
DQ32	36	104	$\overline{E1}$
DQ33	37	105	DQ34
V$_{SS}$	38	106	DQ35
DQ36	39	107	DQ37
DQ38	40	108	V$_{CC}$
DQ39	41	109	DQP4
DQ40	42	110	DQ41
V$_{CC}$	43	111	DQ42
DQ43	44	112	DQ44
DQ45	45	113	V$_{SS}$
DQ46	46	114	DQ47
DQP5	47	115	DQ48
V$_{SS}$	48	116	DQ49
K1	49	117	V$_{SS}$
V$_{SS}$	50	118	DQ50
DQ52	51	119	DQ51
DQ53	52	120	DQ54
DQ55	53	121	DQ56
DQP6	54	122	V$_{SS}$
V$_{CC}$	55	123	DQ57
DQ58	56	124	DQ59
DQ60	57	125	DQ61
DQ62	58	126	DQ63
DQP7	59	127	V$_{CC}$
A0	60	128	A1
A2	61	129	A3
A4	62	130	A5
A6	63	131	A7
A8	64	132	NC
A10	65	133	A9
A12	66	134	A11
A14	67	135	A13
V$_{SS}$	68	136	A15*

(c)

PIN NAMES	
A0 – A15	Address Inputs
K0, K1	Clock
$\overline{W0}$ – $\overline{W7}$	Byte Write
$\overline{E0}$, $\overline{E1}$	Module Enable
$\overline{G0}$, $\overline{G1}$	Module Output Enable
DQ0 – DQ63	Cache Data Input/Output
DQP0 – DQP7	Data Parity Input/Output
$\overline{TSC}$	Transfer Start Controller
$\overline{TSP}$	Transfer Start Processor
$\overline{BAA}$	Burst Address Advance
PD0 – PD2	Presence Detect
V$_{CC}$ . . . . + 5 V Power Supply
V$_{SS}$	Ground

Figure 9.17   A 136-pin burst SRAM DIMM module: (a) top view; (b) pinout; (c) pin names.

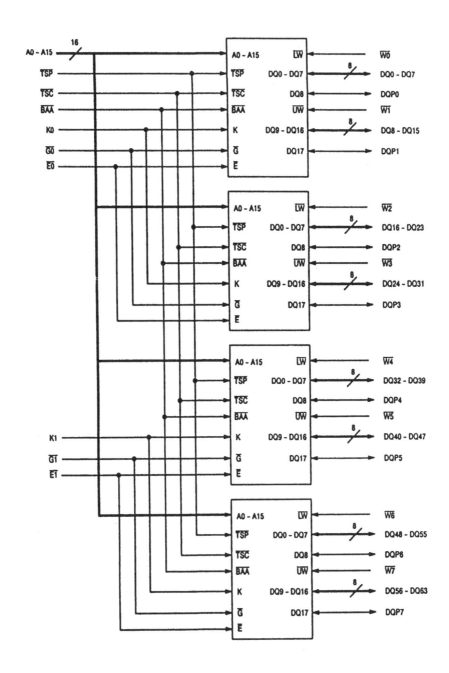

Figure 9.18   Block diagram of a 64K×72 module (four 64K×18 SRAMs)

SRAM chips can also be mounted in a single-chip type package with other system components to improve speed by reducing the length of the wires between these components.

An example of the multichip package is the Intel P6 processor which includes both the processor and an L2 cache consisting of a 256KB SRAM chip which are wire bonded into a single package. The interface between the two is a form of the GTL low swing interface discussed in the last chapter.

The advantage of this configuration is the short transmission lines of the dedicated 64b bus between the processor and SRAM cache chip which permits transfers between the two components at full processor clock speed and width. Clock skew across the processor and between the processor and L2 cache are carefully controlled.

9.5.3   SRAM Multichip Modules

The most common SRAM multichip modules, however, are the bare die SRAMs on a substrate. These can either be mounted face up and wire bonded, or they can have the bonding pads metal bumped and be mounted face downward on a system of interconnects.

Full multichip modules using flipchip technology are one of the most compact systems that can be achieved using separate components. This is because the bonding pads are underneath the chips and no packages are involved. The area occupied by the package is just the area of the bare chip plus the area between chips required for mounting.

The tight geometries of the multichip module assembly also make it capable of the highest speed of any assembly technology.

The advantages and disadvantages of multichip modules are outlined in Table 9.5.

Table 9.5   Advantages and disadvantages of multichip modules

Advantages:
   Small form factor with minimal interchip spacing
   High system speed with reduced wire length and clock skew
   Low EMI radiation due to shortened external wires
   Creative chip combinations

Disadvantages:
   Handling and testing bare die
   Reduced cumulative yield
   Burn-in for bare die
   Warranty for bare die
   Module repair
   Heat and power dissipation

Solutions for disadvantages of multichip modules

Partial solutions for the disadvantages of multichip modules have been found in the areas of module repair, test and burn-in. It is expected that most of these disadvantages can be overcome in time.

The problems of module repair have been lessened by recent innovations of the chip manufacturers. For example, some processors used in modules are built with multiple (duplicate) bond pads. This improves the yield due to wire bond error on the most expensive chip in the module. The substrate can also have multiple bond pads or can have oval bond pads as Micro Module Systems has done. Either method allows more than one wire bond to be attempted.

The multiple bond pads are used when a faulty wire bond occurs. The bond wire can be snipped and rebonded immediately. This eliminates the problem of attempting to remove a chip that has been soldered down or the need to throw away an expensive chip after a single wire bond error [9].

An innovative solution to testing bare die is a test fixture developed by Micro Module Systems that clamps the bare die against a set of test probes on a substrate and acts as a temporary package for the purpose of test or burn-in. It can be used in a hand-held test fixture. While still not a production process, it is a step in the right direction [9].

9.6 Package Considerations in Replacing or Upgrading a Cache SRAM

9.6.1 General Considerations

The decisions made in upgrading a cache for speed or density are similar regardless of the particular densities involved or for that matter the product involved. To a large extent these considerations are related to package, speed, pinout, and board space. To discuss these an example of an upgrade of a 256KB cache original made with 32K×9 Burst SSRAMs will be considered.

A 256KB data cache made with a bank of eight 32K×9 burst SSRAMs can be replaced by four 32K×18 burst SSRAMs, two 32K×32 or 64K×18 burst SSRAMs, or eventually by one 32K×64 burst SSRAM.

Various considerations in upgrading a cache by replacing the SRAMs will be considered by cache type. The replacement alternatives for the 32K×9 in a 256KB system are shown in Table 9.6. They include 32K×18 and 32K×32 for cost reduction and the 64K×18 for density upgrade to 256KB cache.

Table 9.6 Replacement for 32K×9 in 256 KB system

Upgrade: first generation	32K×18, 5 V
Upgrade: second generation	64K×18, 5 V
Redesign: next generation system	32K×32, 3.3 V
	64K×18, 3.3 V

9.6.2   First Generation Upgrades

The 32K×9 burst SRAM can be replaced by the 32K×18 for cost and power reduction in next generation versions of existing systems. A 256KB bank using eight 32K×8(9) SRAMs, for example, can be replaced by four 32K×18 chips.

Simple replacement of eight chips by four requires minimal system modification (change of footprint) and should result in a system cost reduction due to the reduction both in the amount of silicon and in the number of packages.

If the silicon adder for a ×18 over a ×9 is on the order of 20 percent, then silicon is reduced from a normalized factor of eight (for eight chips) to 4.8 (four chips x 1.2) having a 60 percent factor in silicon area and therefore a 40 percent cost reduction in silicon in the same technology.

The package cost is also reduced. For example, a package commonly used for the 32K×9 is the 40-lead PLCC and for the 32K×18 is the 54-lead PLCC. It will be assumed that the total cost comparison between similar packages is a linear factor of the pin count, all other factors being equal. In this case, the package cost reduction is 68 percent in going from the eight 40-lead to four 52-lead packages.

The footprint of a 40-lead PLCC with 50 mil pin pitch is 16×16 mm^2 (256 mm^2) while that of a 52-lead PLCC with the same 50 mil pin pitch is 18.5 × 18.5 mm^2 (342 mm^2). The space on the board is therefore reduced to 67 percent ((4 x 342) / (8 x 256)) if only the space occupied by the package is considered. The reduction is more if the interpackage spacing is taken into account. If a module is used, then there will be no change in board space in the system.

Reducing chip count also reduces power. A 5 V CMOS 32K×9, for example, is rated by one manufacturer at 1 W and a 32K×18 at 1.5 W, so maximum power, in this case, goes from 8 W to 6 W.

In summary, the advantages of replacing the 32K×9 by the 32K×18 are cost reduction, chip count reduction, and power reduction which affects speed.

The chip count reduction from eight chips to four reduces the silicon area by 60 percent, the package cost by 68 percent, and the board space by 74 percent. Power is reduced from a maximum of 8 W to 6 W. The smaller number of chips reduces the length of wiring in the system and enhances the speed.

9.6.3   Second Generation Upgrades

Second generation upgrades of 256KB systems can go to the 64K×18(16). A simple upgrade in cache capacity of the same system from 256KB to 512KB can be made by replacing four 32K×18 5 V chips with four 64K×18 5 V chips. The package is the same so the footprint is the same, as is the power dissipation.

It is not clear whether this would be a major application since the cache hit rate would not be significantly improved in going from a direct mapped 256KB cache to a direct mapped 512KB cache. If done, it will be more for the marketing value of the perceived performance improvement than for the actual system improvement.

9.6.4 *Next Generation System Redesigns*

Next generation systems, meaning that there has been a system redesign, might use the 32K×32 3.3 V. The 32K×32 offers the same benefit of reduced package count and reduced power consumption that the 32K×18 offered over the 32K×9. However, it reduces the board space in the system by an even larger factor, since the two parts are in the same package and the number of packages is being reduced by a factor of two. It also reduces the cost of the silicon even more since the ×18 part with parity bits is being replaced by a ×32 part without the four parity bits. Finally two 32K×32 burst SSRAMs can be replaced with one 32K×64 burst SSRAM for both cost reduction and board space saving reasons.

Bibliography

1. Samsung, *DRAM Databook*, 1995.
2. Texas Instruments, *MOS Memory Databook*, 1995.
3. IBM, *DRAM Module Databook*, 1995.
4. NEC Memory Databook.
5. Hitachi, *4M×72 Synchornous DRAM Module, Target Spec*, Rev. 0, 4 November, 1994.
6. Motorola, *Fast Static RAM Databook*, 1994.
7. Micron Technology, *DRAM Databook*, 1995.
8. Memory Strategies International, High Density Cache SSRAM Report, 1995.
9. Micro-Module Systems, Private Correspondence.
10. Motorola, Fast Static RAM Databook, 1995.
11. Toshiba PC100, 168-Pin DIMM Module, THM641661BEG, February 24, 1998.
12. SLDRAM datasheet, SLDRAM Consortium, 1998.

Index

Printed and bound by CPI Group (UK) Ltd, Croydon, CR0 4YY

27/10/2024

14580216-0005